Bringing Math Students Into the Formative Assessment Equation

To Joan Ferrini-Mundy and Richard Schwab, two very influential teachers and mentors who helped set me on this path, and shaped my approach to the teaching profession in so many valuable ways: with appreciation and gratitude. —SJC

To my late mother, who encouraged me to do what I loved and to be who I am. —EK

To the FACETS teachers, whose energy, dedication, and thoughtfulness provided rich experiences for us to learn from. —SJC, EF, EK, CT

Bringing Math Students Into the Formative Assessment Equation

Tools and Strategies for the Middle Grades

Susan Janssen Creighton

Cheryl Rose Tobey

Eric Karnowski

Emily R. Fagan

A SAGE Company

FOR INFORMATION:

Corwin

A SAGE Company

2455 Teller Road

Thousand Oaks, California 91320

(800) 233-9936

www.corwin.com

SAGE Publications Ltd.

1 Oliver's Yard

55 City Road

London EC1Y 1SP

United Kingdom

SAGE Publications India Pvt. Ltd.

B 1/I 1 Mohan Cooperative Industrial Area

Mathura Road, New Delhi 110 044

India

SAGE Publications Asia-Pacific Pte. Ltd.

3 Church Street

#10-04 Samsung Hub

Singapore 049483

Acquisitions Editors: Robin Najar and Erin Null

Associate Editor: Desirée A. Bartlett

Editorial Assistant: Andrew Olson

Production Editor: Veronica Stapleton Hooper

Copy Editor: Michelle Ponce

Typesetter: C&M Digitals (P) Ltd.

Proofreader: Wendy Jo Dymond

Indexer: Sheila Bodell

Cover Designer: Candice Harman

Marketing Manager: Amanda Boudria

Printed in the United States of America.

A catalog record of this book is available from the Library of Congress.

ISBN: 978-1-4833-5010-3

This book is printed on acid-free paper.

15 16 17 18 19 10 9 8 7 6 5 4 3 2 1

Contents

Preface   ix

Acknowledgments   xv

About the Authors   xvii

**Chapter 1: Using Formative Assessment to
Build Student Engagement in Mathematics Learning**   **1**

  Being a Self-Regulating Learner   3

  Using Formative Assessment Practices to Develop
    Self-Regulation Skills   5

    What Is Formative Assessment?   5

    Overview of the Aspects of Formative
      Assessment   7

  Using Formative Assessment in Your Classroom   12

    The Formative Assessment Cycle   12

    Putting Your Students Front and Center   15

    Comparing Formative Assessment Practices
      to Your Teaching Practices   16

  Teaching Students How to Participate in
    Formative Assessment   18

    Resources for Teaching Students About
      Formative Assessment   19

  How to Use this Book to Learn What *You* Want to Learn   20

  Conclusion   21

  Resources   22

    Learning Resources   22

    Reference Resources   22

**Chapter 2: Using Mathematics Learning
Intentions and Success Criteria**   **23**

  What Are Learning Intentions and Success Criteria?   24

    Characteristics of Learning Intentions and Success Criteria   27

  Creating Learning Intentions and Success Criteria
    for Your Classroom   36

    Your Role Before the Lesson: Planning With
      a Learning Intention and Success Criteria in Mind   38

    Your Role During the Lesson: Sharing and
      Using Learning Intentions and Success Criteria   53

  Conclusion   58

    Recommendations in This Chapter   59

Resources   59
   Learning Resources   60
   Reference Resources   60
   Planning Resources   61
   Classroom Resources   61
   Classroom Materials   62

Chapter 3: Gathering, Interpreting, and Acting on Evidence   63
What Is Evidence?   64
   Sources of Evidence   68
   Aligning Evidence to Learning Intentions
     and Success Criteria   69
Using Evidence in Your Classroom: The Teacher's Role   71
   Eliciting Evidence of Meeting the Success Criteria   72
   Interpreting Evidence of Meeting the Success Criteria 86
   Choosing a Responsive Action   90
   Managing the Information   100
The Student's Role and How You Can Develop and Support It   102
   Participation   102
   Self-Monitoring   103
Conclusion   105
   Recommendations in This Chapter   105
Resources   106
   Learning Resources   106
   Reference Resources   107
   Planning Resources   107
   Classroom Resources   107
   Classroom Materials   108

Chapter 4: Providing and Using Formative Feedback   109
What Is "Formative Feedback"?   110
   Characteristics of Formative Feedback   115
   Whole-Group Feedback Versus Individual Feedback   123
Using Formative Feedback in Your Classroom   128
   Feedback or Instruction?   128
   The Teacher's Role Before the Lesson: Planning
     for Formative Feedback   130
   The Teacher's Role During the Lesson   134
   The Student's Role and How You Can
     Develop and Support It   143
Conclusion   148
   Recommendations in This Chapter   149
Resources   149
   Learning Resources   150
   Reference Resources   150
   Planning Resources   150
   Classroom Resources   150
   Classroom Materials   151

**Chapter 5: Developing Student Ownership
and Involvement in Your Students   153**
Student Ownership and Involvement   154
What Do Students Need to Learn?   156

The First Question: Learning to Use Learning
    Intentions and Success Criteria to Clarify Goals   158
The Second Question: Learning to Use Success Criteria,
    Evidence, and Formative Feedback to Self-Assess   160
The Third Question: Learning to Determine Next Steps   162
The Final Step: Taking Action   165
Helping Students Develop Ownership and Involvement in
    Their Mathematics Learning   166
    Helping Students Learn About Learning
        Intentions and Success Criteria   167
    Helping Students Learn About Evidence   175
    Helping Students Learn About Formative Feedback   178
    Helping Students Learn About Taking Action   181
Self-Regulation: It's Not a Linear Process   183
Conclusion   187
    Recommendations in This Chapter   188
Resources   189
    Learning Resources   189
    Reference Resources   189
    Planning Resources   189
    Classroom Resources   189
    Classroom Materials   190

Chapter 6: Using Mathematics Learning Progressions   **191**
What Is a Learning Progression?   192
    Characteristics of Learning Progressions   195
How Can You Use Learning Progressions in Your Instruction?   198
    Purpose 1: Using Learning Progressions to Provide
        Mathematics Background Information   198
    Purpose 2: Using Learning Progressions to Plan Instruction   204
    Purpose 3: Using Learning Progressions
        to Help You Diagnose Student Difficulties   210
Using Learning Progressions to Help Develop Student
    Ownership and Involvement   213
Conclusion   214
    Recommendations in This Chapter   215
Resources   215
    Learning Resources   215
    Reference Resources   216
    Planning Resources   216

Chapter 7: Establishing a Classroom Environment   **217**
Elements of the Classroom Environment   218
The Social and Cultural Environment: Promoting Intellectual
    Safety and Curiosity   222
    Intellectual Safety and Risk Taking   223
    Intellectual Curiosity   229
    Putting Them All Together   233
The Instructional Environment: Framing Instruction to
    Encourage and Make Visible Students' Thinking and to
    Optimize Learning   234
    Math Tasks   234

Students' Responsibility for Doing the Learning 236
The Physical Environment: Keeping Resources Available 237
Learning Materials 240
Evidence-Gathering Materials 241
Self-Assessment Materials 241
Peers as Resources 242
Conclusion 243
Recommendations in This Chapter 244
Resources 244
Learning Resources 244
Reference Resources 245
Classroom Materials 245

Chapter 8: Moving Toward Implementation 247
Implementation Principles for Formative Assessment 249
Learning Progressions 249
Learning Intentions and Success Criteria 249
Eliciting and Interpreting Evidence 250
Formative Feedback 251
Student Ownership and Involvement 251
Classroom Environment 253
Sustaining Your Effort Over the Long Term 254
Conclusion: Final Words of Encouragement 267
Recommendations in This Chapter 268
Resources 268
Learning Resources 268
Reference Resources 269
Planning Resources 269

Appendix A: Resources 271

**Appendix B: Implementation Indicators
for Formative Assessment** 277

References 283

Index 285

Additional resources for *Bringing Math Students Into the Formative Assessment Equation: Tools and Strategies for the Middle Grades* can be found at Resources.Corwin.com/ CreightonMathFormativeAssessment

Preface

In 2008, we saw an increase in attention to formative assessment yielding interesting work: excellent books being published, such as those by Black and Wiliam, Lee, and Bright and Joyner; informative articles such as those by Wylie and Heritage for the Council of Chief State School Officers (CCSSO); and district and statewide initiatives being built, such as those in Iowa and Syracuse, New York. At the same time, with only a few exceptions (such as Bright and Joyner's, 2004, *Dynamic Classroom Assessment*), these materials and initiatives cut across school content, using examples from science, language arts, and social studies as well as mathematics. Our experiences with professional development for teachers led us to believe that while these cross-content descriptions are invaluable to the field, a focus on what formative assessment looks like in the mathematics classroom is needed for teachers of math. We often hear frustration from cross-content professional development, as teachers say, "I see how this works in the example subject, but I don't see how it fits in mine." (This is true whether, for example, the teacher's subject is math and the example is language arts, or *vice versa*.) As you'll see in this book, effective implementation of formative assessment is deeply tied to the subject content, so providing examples and suggestions specifically for mathematics teachers is paramount.

To respond to this need, we have written this book for middle grades (5–8) teachers of mathematics, including special education teachers. While much of the information within is transferable to other grades, and even to other content, we maintain our focus on mathematics at the Grade 5–8 level to make this book as relevant to this audience as possible.

This book provides

- an extensive description of formative assessment, including examples from middle grades mathematics classes to focus support for math teachers in implementing formative assessment in ways that a subject-general book cannot, and
- concrete tools, strategies, and resources, informed by research and field-tested by teachers, to
  - support your implementation of various formative assessment practices: setting and using learning goals and criteria for success, eliciting and interpreting evidence, and acting on the information; and

○ make explicit to students their role in the formative assessment process: understanding the learning goal; self-monitoring, self-assessment and reflection; and using peers as resources to provide and act on feedback.

Through the professional development work and associated research that helped us to produce this book (see "Our Work"), the ideas in this book have been tested. We are proud to have included voices of teachers throughout the book, as we have learned much from the teachers who have participated in and helped to improve and refine our work by using in it their classrooms. We hope that hearing from them helps you to bring the ideas into practice.

This book is rich with information and resources that you will find useful for many years. We hope you will not read it once and put it back down—rather, we hope that you will return to it again and again, as you deepen your understanding and move forward to integrate formative assessment into your daily practices.

■ OUR WORK

With a grant from the National Science Foundation (Grant # DRL-0918438), we began work to develop *Formative Assessment in the Mathematics Classroom: Engaging Teachers and Students* (or FACETS), a program and resources to introduce middle school mathematics teachers to formative assessment and to support them as they began to implement formative assessment practices in their classrooms. We based our initial work on that of many educational researchers and teacher educators and on our own prior work in teacher professional development, formative assessment, and mathematics curriculum development and teacher training. After a year of initial development, we worked with one group of teachers for 2 years, learning about how they made sense of the information we were providing.

As we implemented our initial program with our first cohort of teachers, the FACETS research team gathered and interpreted data to study teacher's learning of formative assessment. We worked with a second cohort of teachers for another 2 years, again learning even more about what worked and what didn't.

The researchers provided significant input to help us shape the program, guided by the questions:

• How do middle grades teachers make sense of and learn to implement formative assessment practices?
• What barriers and challenges do they face in implementing formative assessment?
• What are supports that can help them move through these challenges?

For example, we learned that

• envisioning students as the primary consumers of formative assessment information in teacher's conception of the formative assessment process early made even the initial work more meaningful and powerful.

- teacher learning of formative assessment takes time. As you'll see, formative assessment has many pieces to it; teachers need time to get to know each piece individually and then more time to put those pieces together.

Figure 0.1 illustrates, in a simplified way, how teachers were able to approach learning about formative assessment. As they were becoming familiar with formative assessment and considering how to manage the different pieces, they were mostly focused on their own actions, though the student role in formative assessment did get some attention. Later, they were able to switch their focus to be more on the student, as they worked toward full integration of the pieces within their teaching practices.

This book, and the companion resource website (Resources.Corwin .com/CreightonMathFormativeAssessment), are the results of that work.

HOW TO USE THIS BOOK ■

We've written this book to be read through from beginning to end, though we recognize that different people will want different things from a resource like this. If you decide not to read each chapter in order, we highly recommend at least reading Chapters 1 and 2 first. (Depending on how you plan to bring this information into your classroom, Chapter 8 might also be a good one to read after that—see the last paragraph in this

Figure 0.1 A Simplified View of Teacher Learning of Formative Assessment

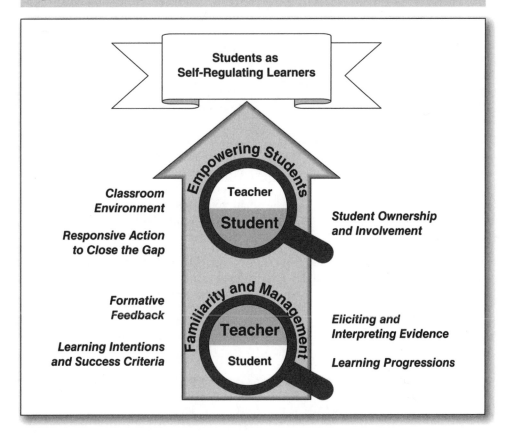

section.) Given our emphasis on student involvement and self-regulation, Chapter 5 may seem like an important chapter to read early, however, we believe it will make most sense *after* you have read Chapters 1 through 4. Those chapters include some previews of what you will read in Chapter 5, so you will still be getting some idea of the student role in the formative assessment process throughout the early chapters.

Resources, available at Resources.Corwin.com/CreightonMathFormativeAssessment, are highlighted throughout the book. When you see a reference to a resource, you may want to stop reading to look—we support that desire, but with a few exceptions, this isn't expected. For those exceptions, such as a couple of places in Chapter 1, we explicitly request that you take a moment to look over or work with a particular resource before continuing. In those cases, we are trying to re-create some popular and effective professional development activities that we think will help you develop some initial understanding of formative assessment. Of course, you may choose to continue reading rather than stop. You know how you learn better than we do!

Two of the following icons accompany each resource, both in this book and on the companion resource website. One icon represents what type of resource it is (interactive, poster, etc.), and the other represents how it's meant to be used.

Usage Icons		Type Icons	
	Reference Resource: Information that you, the teacher, may want to come back to.		**Summary Card:** Reference cards that summarize important information. We recommend printing these on sturdy paper or laminating and punching a hole in one corner so you can put them on a key ring to keep them together.
	Learning Resource: Intended to be used by you, the teacher, as part of your own learning about formative assessment.		**Interactive:** Web-based interactive pages, accessible through the companion resource website.
	Planning Resource: Intended to be used by you, the teacher, to help you in your lesson planning process. You may want to use these on an ongoing basis.		**Poster:** Large format PDF files that you can print, or have printed at a commercial office printer, to be hung in your classroom and used to help students with their part in formative assessment practices.
	Classroom Resource: Lesson plans and strategies for you to use with your students.		**Video:** Video that provides greater insight into teachers' thoughts and practices in the classroom.
	Classroom Material: Intended to be used primarily by the students.		**PDF:** Resources that can be printed on letter- or legal-sized paper.

Finally, most chapters include several recommendations to help you integrate formative assessment into your classroom practices; we suggest you give yourself time to integrate them gradually—choose a few to focus on until they become familiar, and then you can choose another small group when you're ready to move on. Chapter 8 provides some specific recommendations on how to do this gradual implementation, so if you prefer to begin implementing before you've completed the whole book, we recommend looking at that chapter—and then return to Chapter 8 often as you progress.

Acknowledgments

We would like to acknowledge many people who have helped and supported us in creating this book.

First, a very grateful *thank you* to the teachers who worked with us, giving us their time and attention to provide feedback on their experiences: Nancy Adamoyurka, Sara Allegretti, Bekah Aucoin, Lisa Beede, Maryanna Biedermann, Laurie Boosahda, Susan Brass, Nancy Cachat, Debra Chamberlain, Jan Chamberlain, Sara Churchill, Terri Clark, Laura DeSimone, Jessica Dineen, Nancy Dobbins, Joelle Drake, Michelle Eastman, Eli Edinson, Kerri Falcone, Dory Fish, Carolyn Flanders, Sheri Flecca, Judy Forrest, Josh Fox, Maura Ghio, Linda Gloski, Megan Goff, Angie Goldberg, Meredith Gonzalez, Michele Greco, Maureen Griffin, Tracey Hartnett, Renée Henry, Pat Henyan, Sharon Johnson, Dan Johnston, Patrick Kelcourse, Tracy Kinney, E. J. Kluge, Marguerite Lackard, Tami LaFleur, Kathleen Lam, Jesse Latimer, Jane Lewandowski, Errol Libby, Tom Light, Jessica Lloyd, Megan Maguire, Mandy Marrella, Laura McDuff, Guy Meader, Judy Morgan, Jonathan Newman, Michele O'Connor, Lauren O'Malley, Patricia Paul, Nancy Philbrick, Joan Savage, Michelle Schechter, Ricki Scheeder, Nicole Simkins, Diana Smith, Mike Soucy, Wendy Stebbins, Michele Torkomian, Bridget Wade, Nichole Walden, Alison Washington, Mae Waugh, Adam West, Jessica West, Ruth Wilson, Ed Worcester, Yan Yii, and Steve Zakon-Anderson. We also would like to thank the following educators who embraced many of these ideas during a schoolwide effort to implement formative assessment: Amanda Bolanda, Karla Bracy, Amy Bru, Laura Cummings, Peggy Dorf, Terri Eckes, Laura Foley, Georgina Grenier, Melissa Guerrette, Kelly Langbehn, Lauren LaPointe, Susan Leunig, Cathleen Maxfield, Nathan Merrill, Stephanie Pacanza, Roberta Polland, Cynthia Powers, Kim Ramharter, Sara Roderick, Joann Smith, Gal Stetson, Lisa Stevens, and Kimberly Tucker.

We are most grateful for the support, sacrifice, and patience shown by our families throughout the work involved in bringing this book to fruition: Doug Creighton, Evan Janssen, Alison Janssen; Mark Saperstein; Corey Tobey, Robert Spadea, Carly Rose, Jimmy Rose, Bobby Rose, Samantha Tobey, Jack Tobey; Sean Fagan, Nellie Fagan, and Seamus Fagan.

Next, our colleagues at Education Development Center (EDC): Lynn Goldsmith and Sophia Mansori formed the research team for FACETS, and Cynthia Char (of Char Associates) served as our external evaluator. We would have been lost without our administrative assistants, Mari Halladay, Carlene Kaler, and Michelle Raymond. Our former colleague,

Fred Gross, was instrumental in assembling the team. And we enjoyed and appreciated the support of many others in our EDC community.

We also want to thank the advisors to the FACETS project: Colleen Anderson, David Baumgold, Steve Benson, Paul Black, Mark Driscoll, Margaret Heritage, Karen King, Kit Norris, Paola Sztajn, and Nancy Zarach.

As we drafted this book, we received wonderful feedback from several reviewers: Carolyn Arline, George Bright, Peg Brown, Nicole Clark, Deb Cook, Heather Daigle, Michelle Eastman, Matthew Lunt, Diana Smith, Joan Taczli, and Ruth Wilson.

We want to extend a special thank you to the folks at Corwin who worked diligently to help us bring you this book, and the companion resource website: Desirée Bartlett, Arnis Burvikovs, Robin Najar, Andrew Olson, Michelle Ponce, Ariel Price, Maura Sullivan, and Veronica Stapleton Hooper.

Finally, as coauthors, we want to acknowledge each other. Working together over the last 5 years to create and evolve the ideas, tools, and strategies that are the foundation of this resource has been a truly collaborative experience.

PUBLISHER'S ACKNOWLEDGMENTS

Corwin would like to thank the following individuals for taking the time to provide their editorial insight and advice:

Lyneille Meza
Coordinator of Data & Assessment
Denton ISD
Denton, TX

Dr. Marc Simmons
Principal
Ilwaco Middle School
Long Beach, WA

Rita Tellez
Math Coordinator
Ysleta Independent School District
El Paso, TX

Morris White
Secondary Math Teacher
Alamosa High School
Alamosa, CO

About the Authors

Susan Janssen Creighton is a senior mathematics associate at Education Development Center (EDC) in Massachusetts. She has worked in mathematics education for 30 years, both in schools and at EDC, where her work has focused largely on K–12 mathematics curriculum development and mathematics teacher professional development. Currently, her work focuses on helping mathematics teachers adopt and successfully implement formative assessment practices and on supporting teachers' understanding and use of the CCSS Standards for Mathematical Practice. As a member of the National Science Foundation (NSF)–funded project, *Formative Assessment in Mathematics Classrooms: Engaging Teachers and Students* (FACETS), she was a lead facilitator for several of the participating districts.

Creighton has written print and online materials for numerous clients, including the international Department of Defense schools, the Columbus, Ohio, public schools, the National Board of Professional Teaching Standards, Everyday Learning publishers, the PBS *TeacherLine* project, the Massachusetts Department of Education, and the *E-Learning for Educators* project funded by the U.S. Department of Education. She has also served as the director of the *MathScape* Curriculum Center, a national center that supported the implementation of the NSF-funded mathematics curriculum *MathScape* developed at EDC, for which she was also one of the writers, and has led numerous teacher professional development opportunities for middle and high school teachers on the teaching and learning of mathematics. Prior to coming to EDC, she taught middle school and high school mathematics for a number of years in Portland and Saco, Maine and in Brookline, Massachusetts. She received a BA in mathematics and a MEd in Secondary Education, with a concentration in curriculum, both from the University of New Hampshire. She currently lives in western Massachusetts with her husband, her two teenagers, and the world's softest dog.

Cheryl Rose Tobey is a senior mathematics associate at EDC in Massachusetts. She is the project director for *Formative Assessment in the Mathematics Classroom: Engaging Teachers and Students* (FACETS) and a mathematics specialist for *Differentiated Professional Development: Building Mathematics Knowledge for Teaching Struggling Students* (DPD); both projects are funded by the NSF. She also serves as a director of development for an Institute for Educational Science (IES) project, *Eliciting Mathematics Misconceptions* (EM2). Her work is primarily in the areas of formative assessment and professional development.

Prior to joining EDC, Tobey was the senior program director for mathematics at the Maine Mathematics and Science Alliance (MMSA), where she served as the coprincipal investigator of the mathematics section of the NSF-funded Curriculum Topic Study, and principal investigator and project director of two Title IIa state Mathematics and Science Partnership projects. Prior to working on these projects, Tobey was the coprincipal investigator and project director for MMSA's NSF-funded Local Systemic Change Initiative, Broadening Educational Access to Mathematics in Maine (BEAMM), and she was a fellow in Cohort 4 of the National Academy for Science and Mathematics Education Leadership. She is the coauthor of six published Corwin books, including seven books in the *Uncovering Student Thinking* series (2007, 2009, 2011, 2013, 2014), two *Mathematics Curriculum Topic Study* resources (2006, 2012), and *Mathematics Formative Assessment: 75 Practical Strategies for Linking Assessment, Instruction, and Learning* (2011). Before joining MMSA in 2001 to begin working with teachers, Tobey was a high school and middle school mathematics educator for 10 years. She received her BS in secondary mathematics education from the University of Maine at Farmington and her MEd from City University in Seattle. She currently lives in Maine with her husband and blended family of five children.

Eric Karnowski is a senior mathematics associate EDC in Massachusetts. He has worked in mathematics education for 25 years, initially as a teacher, then as a textbook editor, and finally as a curriculum developer and teacher professional development provider. Since joining EDC, he has directed the development of the K–5 program *Think Math!* and written numerous activities for the award-winning *Problems with a Point* website. He directed projects to develop several online teacher professional development courses for PBS *TeacherLine*, Louisiana Algebra 1 Online Professional Development, and most recently, the National Board of Professional Teaching Standards in both mathematics and science. In addition, he was a contributing author on *Ways to Think about Mathematics* and the *MathScape* curriculum.

Prior to joining EDC, Karnowski had the distinct privilege to edit influential secondary textbooks for Janson Publications and Everyday Learning, including *Contemporary Mathematics in Context* by the Core-Plus Mathematics Project, *Contemporary Calculus* by the North Carolina School of Science and Mathematics, and *Impact Mathematics* by EDC. He received a BS in Liberal Arts (honors mathematics) and an MS in Mathematics, both from the University of Tennessee, Knoxville. He currently lives in Boston with his husband, Mark, and two large cats, Endora and Tabitha.

Emily R. Fagan is a senior curriculum design associate at EDC in Massachusetts where she has developed print and online curricula as well as professional development and assessment materials in mathematics for 14 years. She was director of the MathScape Curriculum Center, a project funded by the NSF to support schools, districts, and teachers in curriculum implementation, and she directed the revision of *MathScape: Seeing and Thinking Mathematically* (McGraw-Hill, 2005). She was a developer and facilitator of three NSF-funded projects, *Addressing Accessibility in Mathematics* and *Differentiated Professional Development: Building Mathematics Knowledge for Teaching Struggling Students* (DPD) aimed at supporting struggling math learners, particularly those with learning disabilities, and *Formative Assessment in the Mathematics Classroom: Engaging Teachers and Students* (FACETS) the inspiration for this book.

Fagan is the coauthor of two books: *Uncovering Student Thinking About Mathematics in the Common Core, Grades K–2* (2013) and its companion for Grades 3 through 5, as well as book chapters and articles about curriculum implementation and instruction. Prior to joining EDC, Emily taught high school and middle school mathematics in Philadelphia and in Salem and Brookline, Massachusetts. She was a mentor teacher, math coach, and member of the Massachusetts faculty of the Coalition of Essential Schools. She has long been interested in accessibility in mathematics education and improving opportunities for all students to learn and love math. While mathematics has been her focus for the last 2 decades, she has also taught science, social studies, and Spanish. Fagan holds an AB cum laude from Harvard University. She lives in Sudbury, Massachusetts, with her husband and their two children.

1

Using Formative Assessment to Build Student Engagement in Mathematics Learning

This book sets out to answer, in a very concrete way, the question of "What can I do to more fully engage my students in their mathematics learning, even when—or perhaps, *especially* when—they haven't always experienced a lot of success?" For some students, being engaged in math class feels rewarding; they have frequently been successful with learning mathematics and come to class with an internal dialogue that says, "I can do this," "I enjoy learning mathematics," and "I'm confident I can learn new mathematics ideas." However, many other students come to class with a more negative internal dialogue that says, "I can't do this," "I need the teacher to help me know what to do, so I'll just wait until he or she can give me individual help," and, hoping to remain invisible, "*Please* don't call on me!" Those students who feel successful with mathematics and engage readily have learned to do a collection of things, explicitly from a teacher or on their own, that contribute to this feeling of success. As a mathematics teacher, you can help all your students learn to do those same things.

So what are "those same things?" Our list includes the following:

- Students reference the learning goal throughout a lesson to help clarify for themselves what they're learning and where to focus their attention or to provide feedback to others about mathematics work.
- Students have an image of what successful work looks like, either from criteria provided by the teacher or from criteria they have for themselves internally. Students pay attention to these criteria to evaluate how well their mathematics thinking and their work meet the learning goal.
- Students share their mathematical reasoning with you—their teacher— or with peers in small groups or in the whole class. This sharing goes beyond merely listing what steps they followed, to talking about *why* they followed certain lines of mathematical reasoning. Students also are willing to take risks to share their thinking-in-process, even if it is not fully formed or may be inaccurate.
- Students proactively use the resources they have available to them.
- Students pay attention to the feedback they receive, and they use it to revise their work to better meet the learning goal.
- Students work with peers, either thinking through a mathematics task together, helping each other better understand an idea, or giving each other feedback to improve work.

This list of student actions helps provide some specific examples of what we refer to as students becoming *self-regulating learners*, and we say more about that shortly. The research on formative assessment points to a collection of instructional techniques whose potential impact reach far beyond the use of diagnostic tools to assess students' understanding and make instructional in-the-moment adjustments and points to the development of stronger self-regulation in students (Black & Wiliam, 1998; Heritage, 2010; Popham, 2008; Stiggins, 2007). This list may sound idealistic and beyond the scope of many of your students. However, this book results from several years of intensive work in classrooms with real teachers and students who were able to make this a reality.

The collective voice of "we" who are talking to you throughout this book is an author team who has first and foremost been mathematics teachers. We are also professional developers, coaches, and curriculum developers, and we have worked closely alongside colleagues who are researchers. We have spent numerous hours in classrooms with our own students and with other people's students. And we are indebted to a dedicated group of mathematics and special education teachers who worked with us over 5 years as part of the National Science Foundation–funded project whose work is captured in this book. (See the Preface for more information about this project.) The situations portrayed in this book have happened in real middle grades classrooms that are not out of the ordinary, and in many cases, are less than ideal. The voices of teachers that appear throughout this book are the voices of the teachers who worked closely with us and who have agreed to share their words because they wanted others to experience what they learned from this experience. You may encounter some things you will read here that will make you wonder whether this could happen in your classroom, and the teachers' voices are here to tell you that it can.

The work of our project focused on the use of what we call *formative assessment practices* as a vehicle for helping students learn to become more self-regulating in regard to their mathematics learning. Simply put, these practices are the instructional techniques you can use to implement formative assessment in your classroom. While many of the teachers in our project were familiar with an assortment of formative assessment practices and related tools, many of them had not yet learned to realize the full potential of those practices and tools. This book shares with you what they—and we—have learned. We use the rest of this book to flesh out what the bullets in that list of student actions mean and what you can do to help your students get there. The good news is that you are probably already doing many things in your classroom that support students in learning to do these various actions. Our intent is to help you get more mileage out of those things and to help your students learn to participate in their mathematics class in a powerful way.

BEING A SELF-REGULATING LEARNER ■

The process of students becoming self-regulating learners can be described more broadly as students internalizing a set of questions and resulting actions that guide their learning. Figure 1.1 lays out these questions and actions.

Figure 1.1 Thinking Like a Self-Regulating Learner

- What goals am I aiming for (in my learning)?
- Where am I currently in relation to those goals?
- If I have not met the goals, what do I need to do next to be able to meet them?
- I take action to move my learning toward the goals.

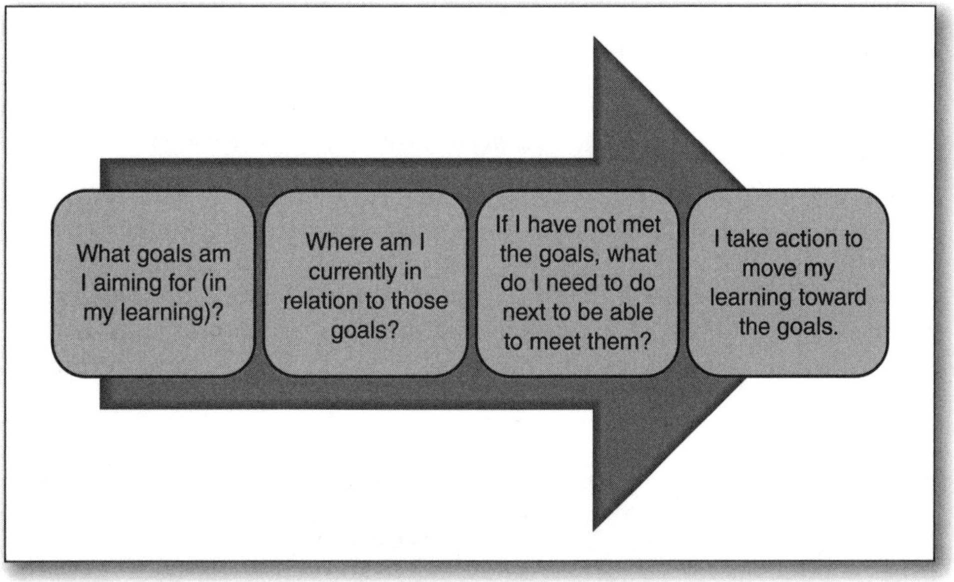

These questions grow out of the work of Royce Sadler, an Australian educator and researcher who described the necessary conditions for self-regulation as follows:

> *[The] learner has to (a) possess a concept of the standard (or goal, or reference level) being aimed for, (b) compare the actual (or current) level of performance with the standard, and (c) engage in appropriate action which leads to some closure of the gap. (1989, p. 121)*

Students may not be aware of these questions nor trying deliberately to answer them, but the questions frame the nature of the internal dialogue going on inside a student's head. If we were to look inside the heads of students, the internal dialogue of a student who is self-regulating would sound quite different from a student who is not; see Figure 1.2.

Figure 1.2   Internal Dialog for Self-Regulating and Non–Self-Regulating Students

	Student who is self-regulating:	**Student who is not self-regulating:**
What goals am I aiming for in my learning?	*"My job is to learn something new—the learning goal for today's lesson."*	*"Do the work to my teacher's satisfaction."*
Where am I currently in relation to those goals?	*"I can try to check myself against the criteria for good work. I can also look at any feedback I've received about my work or my thinking."*	*"I don't know; my teacher will tell me."*
If I have not yet met the goals, what do I need to do next to be able to meet them?	*"I may have some ideas of my own. If I don't, or if I'm unclear, I'll talk to my teacher or peers to figure out how to improve my work or adjust my thinking."*	*"I will wait for the teacher to help me."*
I am able to take action to move my learning toward the goals.	*"I can get myself started. If I get stuck, I can refer to my resources or the feedback I got. I can also get help from a peer or my teacher."*	*"Only if the teacher will tell me what to do next, or when to do it, or how to do it."*

Throughout this book, you will find templates, lessons, strategies, and other resources that you can use to structure your classroom, your instruction, and your interactions with students to move them toward becoming self-regulating learners.

 Reference Resource: Go to Resources.Corwin.com/ CreightonMathFormativeAssessment to access this resource.

 Chapter 1 Summary Cards. This printable file provides a small version of the Thinking Like a Self-Regulating Learner diagram. Each chapter has a set of Summary Cards; we recommend printing them on sturdy paper (or laminating them) and punching a hole in one corner so you can keep them on a key ring for easy access.

USING FORMATIVE ASSESSMENT PRACTICES ■ TO DEVELOP SELF-REGULATION SKILLS

Using a collection of formative assessment practices, taken as a whole, can create rich and effective opportunities for all students to learn self-regulation skills. In our work with teachers, we've learned that there are many varied understandings of what *formative assessment* means, so before we go further, it's important to establish how we talk about formative assessment in this book and what these practices are.

What Is Formative Assessment?

In the existing literature, there are many descriptions and definitions of formative assessment. Here are some examples:

- Formative assessment is a planned process in which assessment-elicited evidence of students' status is used by teachers to adjust their ongoing instructional procedures or by students to adjust their current learning tactics. (Popham, 2008, p. 6)
- By providing students and teachers with specific, regular feedback on how well students are mastering key concepts and skills, formative assessment helps teachers create opportunities that maximize the chances of learning happening. (Brookhart, Moss, & Long, 2008)
- [All] those activities undertaken by teachers, and/or by their students, which provide information to be used as feedback to modify the teaching and learning activities in which they are engaged. (Black and Wiliam, 1998, pp. 7–8)

Notice that these definitions have some common elements: They are focused on identifying student success, but they go on to mention the *use* of that information to help students achieve more. They also include recognition of the students' use of the information as well as teacher use.

In our work with teachers, we found a need to establish a common definition that we could rely on. We have adopted the following definition, put forth by the Council of Chief State School Officers (CCSSO):

> *Formative assessment is a process used by teachers and students during instruction that provides feedback to adjust ongoing teaching and learning to improve students' achievement of intended instructional outcomes. (2008, p. 3)*

We chose this definition both because of its characterization of formative assessment as a process and its inclusion of students in that process and because of its comprehensive nature. For example, there are several important pieces to this definition:

- A process: Formative assessment is much more than a type of quiz, test, or other tool that can simply be given to students. Rather, it's a *process* that starts when the teacher begins planning and continues as the teacher and the students are engaged in a lesson and continues even further as the teacher reflects on the lesson and plans the next. We refer to this process throughout the book just as *formative assessment*.

- Used by students and teachers: While the teacher has a major role in implementing formative assessment, the students also have a major role. In fact, Heidi Andrade (2010) characterizes students as the "key producers and consumers of formative assessment information." A significant portion of this book is focused on the student role and how to engage students in it.

- During instruction: The power of formative assessment lies in its timing. Because the process is used *during instruction,* teachers and students are able to adjust and refocus their efforts so that energy is spent where it does the most good, while it is still helpful—that is, while instruction is still underway and while students are engaged with content, rather than after a unit or chapter is done. During instruction allows both for the use of instructional tools or strategies that can provide information in real time during a lesson as well as those that provide information that a teacher can analyze and use after a lesson to plan for the next day.

- Provides feedback to adjust: Formative assessment cannot be formative if the process does not help teachers shape or *form* their instruction based on the real-time information gathered and if it does not help students *form* their learning in response to what is needed. Feedback is one of the mechanisms in formative assessment that helps students learn to adjust their mathematical thinking and their work, and to refocus their efforts if necessary, to better meet the learning goal. Teachers also get feedback from the process so that they can adjust their approach to meet their students' needs.

- Ongoing teaching and learning: It's notable that this definition mentions both teaching and learning, both of which require adjustment as instruction occurs, to respond to the variety of learners in any given classroom. It may be easiest for teachers to pay attention to adjustments in their teaching to try to help students be successful but more challenging or less familiar to pay attention to helping students learn to make adjustments for themselves to their learning. Yet both are important here.

- To improve student achievement of intended instructional outcomes: Of course, improving student achievement is the ultimate goal for all of our efforts to improve our teaching practices. But as we know, students will learn many things from us, and they will not always learn what we intend them to learn! By creating a clear focus on the *intended* instructional outcomes, teachers can avoid energy that is misspent on other unintended outcomes.

This definition provides an excellent description of formative assessment but not the important details—what it looks like and how one implements it in the classroom. Different authors provide a variety of ways to characterize these details; we consolidated many of the views and, through our work with teachers, arrived at a set of common and useful aspects that we consider the *critical* and *supporting* aspects of formative assessment. In the next section, we provide a brief overview of these aspects; the following chapters delve deeper into each.

Before you read that section, though, we'd like to point you to a collection of resources that exist on the website that accompanies this book. These resources include tools you can use in your classroom with students to implement the recommendations you'll encounter throughout this book as well as planning guidelines for your use. In addition, there are a collection of interactive web pages, some of which can be used as planning tools, others of which help illustrate in more depth some of the ideas in this book. At this point, we suggest you take an opportunity to look at one of these interactive pages. Log in to the companion resource website (Resources.Corwin.com/CreightonMathFormativeAssessment) and find the "Formative Assessment Overview" interactive page, which provides several statements about formative assessment by selected notable authors. You can sort these statements and form your own opinion of how they align with each other. You can also see how your choices align with our own characterization of the critical and supporting aspects. This is an activity we've done with numerous teachers in professional development settings, and we have found that it helps provide a clear, big picture of formative assessment in an engaging and informative way.

Learning Resource: Go to Resources.Corwin.com/ CreightonMathFormativeAssessment to access this resource.

Formative Assessment Overview. This interactive web page provides a big picture view of formative assessment.

Overview of the Aspects of Formative Assessment

While there are several different characterizations of the vital elements of formative assessment in various definitions in the field, there are notable similarities among them. In our work with teachers, we focused on these similarities to identify four elements that we call the *critical aspects* of formative assessment. In addition, we identify two *supporting aspects*, which provide intellectual, mathematical, and practical foundations for implementing formative assessment into your daily teaching practices. (You probably saw these six aspects in the Formative Assessment Overview interactive page.) A short description of each is provided here, and you'll see examples of each in the "Using Formative Assessment in Your Classroom" section of this chapter. Chapters 2 through 7 present details about each aspect, including descriptions of the corresponding teacher and student roles within formative assessment as well as recommendations, tools, and strategies to help you implement them in your own instruction.

Critical Aspects of Formative Assessment

- **Learning Intentions and Success Criteria**: This critical aspect requires articulating, and sharing with students, both the learning that you intend to happen in a lesson (the *learning intention*) and the indicators that help both you and your students assess whether that learning is taking place (the *success criteria*). (See Figure 1.3.)

Figure 1.3 Learning Intentions and Success Criteria in Self-Regulation

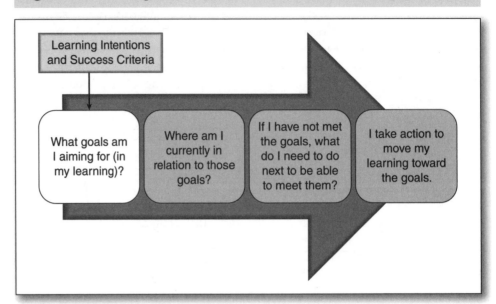

You may be accustomed to writing your lesson goals as student objectives, mastery objectives, "I can . . ." statements, or any one of a number of other formats. Many commonly used formats have favored an emphasis on tangible, observable goals, resulting often in a focus on what students will successfully *do* during the lesson but with less clarity on what students will *understand* mathematically *as a result* of what they do. This understanding is captured in the statement of a learning intention, with the tangible evidence of that learning described in the success criteria. Each of the other critical aspects uses the learning intention and success criteria to focus both your instructional decisions and your students' attention as they try to meet the learning intention.

Whether you name a lesson goal as a *learning intention* or *learning target* or *objective* may just be a matter of semantics. We are choosing to use the language of *learning intention* and *success criteria* here because these are the phrases used in much of the literature on which we based our work with teachers. The importance is not in what you call them but how you articulate them and then use them. There are important characteristics of learning intentions that are effective for implementing formative assessment that may distinguish them from other uses of lesson goals. These characteristics are addressed in Chapter 2, "Using Mathematics Learning Intentions and Success Criteria."

- **Eliciting and Interpreting Evidence**: This aspect entails gathering evidence of student thinking and student skill and interpreting it against the success criteria to determine appropriate next steps. (See Figure 1.4.) All teachers observe what their students do; however, in formative assessment, that gathering of evidence is specifically focused on evidence related to success criteria, and the next instructional steps are similarly focused on using the results to move students' understanding and skills toward meeting the learning intention.

Figure 1.4   Evidence in Self-Regulation

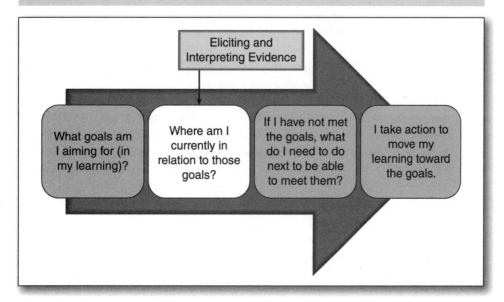

Interpreting evidence of mathematics learning should go beyond answering the question of *can* the students *do* the mathematics, to answering the question of *how* students are *making sense of* the mathematics. Answering these questions requires gathering evidence both about students' fluency with skills and about students' understanding of the underlying mathematical concepts. This aspect is covered in more detail in Chapter 3, "Gathering, Interpreting, and Acting on Evidence."

- **Formative Feedback**: This critical aspect can be a powerful response to student needs, when it's the appropriate next instructional step. There are many kinds of feedback such as motivational feedback,

Figure 1.5   Formative Feedback in Self-Regulation

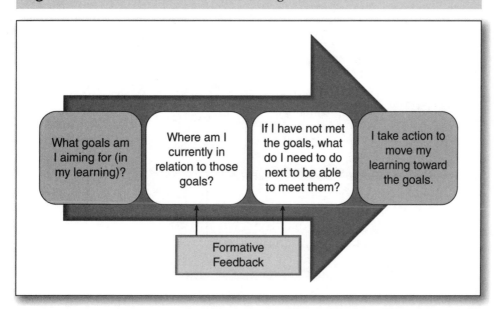

including praise ("You did a terrific job helping your group figure out a solution method!") or suggestions for revision of work ("See if you can find the places where you misplaced the decimal point"). However, feedback that is *formative* refers to the success criteria to provide information that helps the students move farther toward meeting the learning intention. (See Figure 1.5.)

Formative feedback in a mathematics classroom focuses on identifying where students' current mathematics understanding meets or exceeds the goals, where it needs improvement, and what to do next to move their learning forward. Because formative feedback relies so directly on the success criteria, it is difficult to provide a brief example without the surrounding context of the learning intention, success criteria, and the context of the lesson activities. However, Chapter 4, "Providing and Using Formative Feedback," provides more information and numerous examples.

- **Student Ownership and Involvement**: This aspect is listed last, but certainly not least! The goal of helping students learn the skills they need to become self-regulating learners is best viewed as an overarching goal of formative assessment, and as such, makes sense to discuss after the other critical aspects have been introduced. This aspect includes helping students learn to self-assess in relation to the learning intention and success criteria, identify resources (including using each other as peer resources), and take action on their own to move their learning forward. This critical aspect relates to all parts of self-regulation, though self-assessment, identifying resources, and taking action can each be identified with specific different parts of self-regulation. (See Figure 1.6.)

Figure 1.6 Student Ownership and Involvement in Self-Regulation

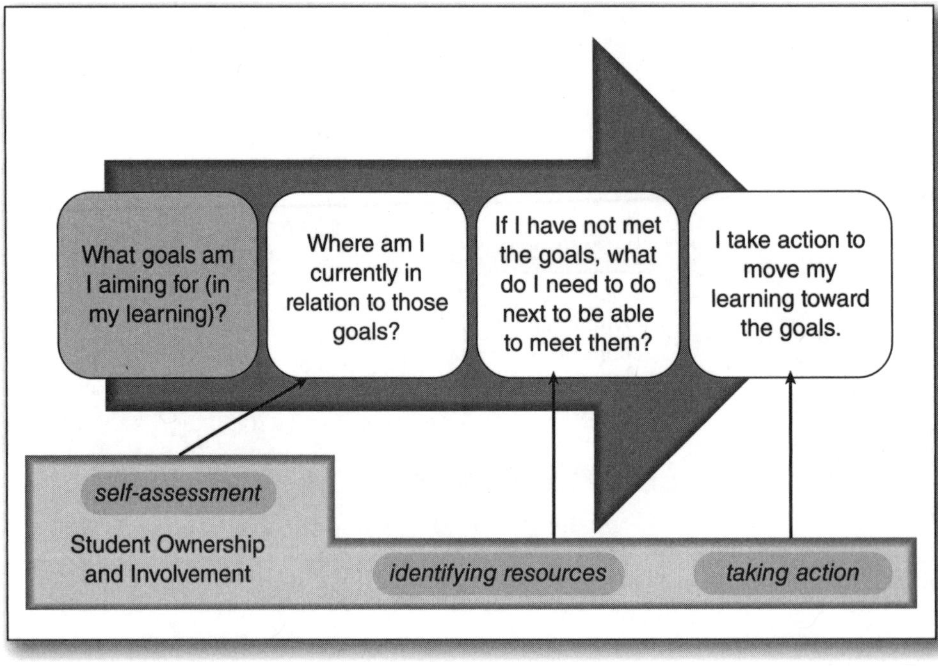

Ideally, an involved student will be looking for the meaning and sense in the key mathematics concepts of the lesson, evaluating his or her work against the success criteria, sharing his or her reasoning with others, giving feedback and responding to feedback from others, and proactively seeking help (from peers, from the teacher, or from other resources such as the textbook or Internet) when needed. While we address student involvement in formative assessment in Chapters 2 through 4, this aspect is considered in depth in Chapter 5, "Developing Student Ownership and Involvement in Your Students."

Supporting Aspects of Formative Assessment

- **Learning Progressions:** A learning progression is an articulation of a pathway through which understanding of content evolves, from basic to more sophisticated understanding. As support for a teacher's use of formative assessment practices, a learning progression provides a context in which a sequence of learning intentions should lie; it also provides insight into gaps in necessary prior knowledge that can contribute to a student's misconceptions or other barriers to learning. Chapter 6, "Using Mathematics Learning Progressions," provides a description and examples of learning progressions and how you might use them to support your own formative assessment practices.

- **Classroom Environment:** Every teacher has a particular classroom environment and knows how fundamental a positive classroom environment is to student learning. However, some environments are more conducive to implementing formative assessment practices than others. Your implementation of formative assessment can be aided or hindered by the physical environment, the cultural and social norms, and the instructional framework of your classroom, so in Chapter 7, "Establishing a Classroom Environment," we discuss these different types of environmental influences.

You might find it helpful to have a summary of these aspects readily available. Throughout the book, we point you to "Summary Card" resources on the companion resource website that summarize information you'll find in this book. We recommend printing these on sturdy paper (or laminating them) and punching a hole in one corner so you can keep them on a key ring for easy access.

Reference Resource: Go to Resources.Corwin.com/ CreightonMathFormativeAssessment to access this resource.

Chapter 1 Summary Cards. This printable file provides a summary of the critical and supporting aspects of formative assessment.

■ USING FORMATIVE ASSESSMENT IN YOUR CLASSROOM

As we delve deeper into each of the critical and supporting aspects in Chapters 2 through 7, we provide specific details about how each appears in the classroom, along with recommendations for planning, implementation, and student use of the formative assessment process. Here, we want to address your role, and your students' role, in more general terms.

The Formative Assessment Cycle

To illustrate how the critical aspects form a coherent process, a lesson-level view is helpful. (In this case, by *lesson* we mean some period of instruction within a unit of study, which may be a single class period or may extend to multiple periods.) We use a diagram we call the Formative Assessment Cycle to illustrate how all the aspects interact within the process. This diagram was informed by the work of Margaret Heritage, Heidi Andrade, Dylan William, and Susan Brookhart, among others, and refined through our work with teachers. Here in Chapter 1, we provide the most basic picture of this cycle (Figure 1.7). As you progress through the chapters, we will gradually provide more detail to flesh out the full version of the cycle.

Our basic view of the Formative Assessment Cycle shows three of our critical aspects, with Eliciting and Interpreting Evidence split into two steps. First, you determine and share the *learning intentions and success*

Figure 1.7   The Evolving Formative Assessment Cycle (Basic View)

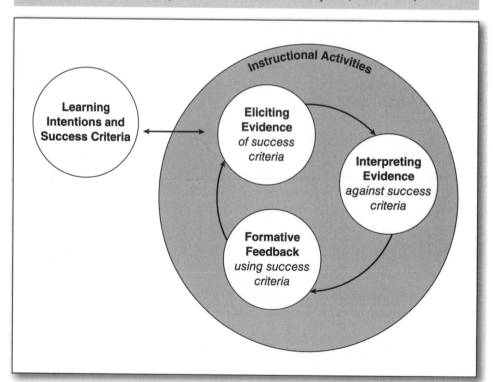

criteria. Then, within the context of the instructional activities of your lesson, are three steps: *elicit evidence* of your students' thinking and learning, focusing on the success criteria; *interpret* that evidence against the success criteria; and then provide *formative feedback* (when appropriate) using the success criteria. At that point, the cycle returns to *eliciting evidence*—did their learning progress?

Example: *For one lesson, Mrs. Jenkins wants students to work toward recognizing how constant change can look in different representations. She decides on this learning intention: "Today, we will learn to relate growth in numeric patterns to growth in corresponding visual patterns." For success criteria, she decides on the following:*

1. *I can describe how a pattern of numbers continues.*

2. *I can describe how a corresponding visual pattern continues.*

3. *I can explain how the way a number pattern grows relates to the way the visual pattern grows.*

For the instructional activities, Ms. Jenkins decides on one Warm-Up and two Exploration Tasks.

- *Warm-Up: Find the missing numbers in this table:*

X	0	1	2	3	4	10	?
Y	0	2	4	?	?	?	28

- *Exploration Tasks: Find the patterns in the sequence of squares and in the sequence of numbers. Be prepared to discuss how the patterns are connected.*

  1. Numbers:

X	1	2	3	4
Y	3	5	7	?

  Figures:

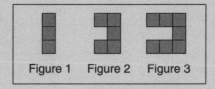

Figure 1   Figure 2   Figure 3

(Continued)

(Continued)

2. *Numbers:*

X	1	2	3	4
Y	3	8	13	?

Figures:

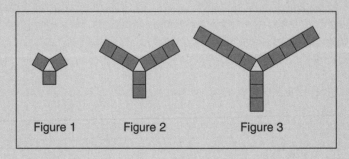

Figure 1   Figure 2   Figure 3

To plan for Exploratory Task 1, she writes out some sample responses that she considers ideal so she can clarify for herself (and later for her students) precisely what meeting her success criteria will look like or sound like. She also plans to bring students together for a discussion that would allow several students to provide their thoughts about relating the numerical and visual patterns. She prepares a few questions:

- *What patterns did you notice in the numbers? In the squares?*

- *How is the way that the number pattern grows like the way that the visual pattern grows?*

- *Do you have a way to describe either pattern in words or in numbers or symbols?*

*After the warm-up activity, Ms. Jenkins reminds them of the problems they worked on the previous day, creating tables of values from a growing visual pattern. She introduces the **learning intention** and explains how this is similar to, and different from, the previous work. She gets volunteers to read **the success criteria**, and then she directs the students to turn to a partner and discuss how success with each criterion shows progress toward the learning intention. After a few minutes, during which Ms. Jenkins listens to several of the discussions, she calls on a few students to share what they talked about, and then she gives them a chance to ask any questions they may have about the learning intention or the criteria.*

*Ms. Jenkins's warm-up required students to complete a table following a simple constant growth pattern. She **gathered evidence** by looking at their responses and **interpreted the evidence** to conclude that most students were able to find the pattern, so she felt confident they could move on to the planned Exploration Tasks. (No need for **formative feedback**.) A few students were unable to articulate how they found their patterns, so she made a mental note to be sure they paired with a student with stronger articulation skills and to check on them during the first task. (This concludes one repetition of the cycle within the instructional activities circle.)*

*As her students complete Exploration Task 1, Ms. Jenkins gathers **evidence elicited** by the task, simply by observing their work and hearing their conversations. She occasionally interrupts a pair to ask for some clarification of their thinking. At the conclusion of the whole-class discussion after Exploration Task 1, which **elicits more evidence**, Ms. Jenkins **interprets the evidence** by comparing their work and conversation to the success criteria. She provides **formative feedback** to the whole class, saying, "I heard some clear explanations of how the number patterns continue and how visual patterns continue—that shows me that we're meeting the first two criteria. However, I'm not sure everyone is clear about our third success criterion, explaining how number pattern growth relates to visual pattern growth. First, can I have a volunteer to tell the class what you think that means?" After a volunteer explains, she gives students a chance to ask questions of the volunteer. (This concludes another repetition of the cycle within the instructional activities circle.)*

Again, this basic view is a simplification of how formative assessment will appear in a classroom. For example, formative feedback might not be an appropriate response to students' needs, as Ms. Jenkins decided after interpreting the evidence from the warm-up. As you dig deeper into each of the aspects and become more familiar with what they mean, we add detail to this cycle and also return to Ms. Jenkins's lesson to see how the added detail is incorporated in her lesson.

Putting Your Students Front and Center

Although we list "student ownership and involvement" as one of our critical aspects of formative assessment, it does not appear in the Formative Assessment Cycle labeled as such, since it really serves as an overarching critical aspect that permeates all the other critical aspects. Students have an important role to play in nearly all aspects of formative assessment. As we add detail to the evolving Formative Assessment Cycle, you will see that we break most of the steps into a teacher role and a student role. (See Figure 1.8.)

Reference Resources: Go to Resources.Corwin.com/ CreightonMathFormativeAssessment to access these resources.

Formative Assessment Cycle. A color version of the completed Formative Assessment Cycle is included in Chapter 8; however, you may want to go to Resources.Corwin.com/Creighton MathFormativeAssessment and take a look now. Look for Chapter 8 Resources.

Formative Assessment Cycle Video. An audiovisual tour of the Formative Assessment Cycle that we are building throughout this book is also included in the Chapter 8 Resources.

Figure 1.8   Example of Teacher and Student Roles in the Formative Assessment Cycle

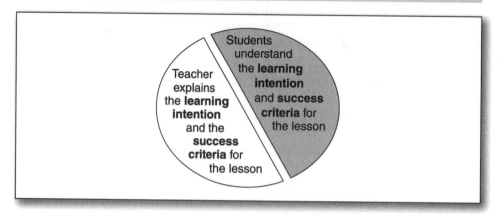

The student is the ultimate consumer of the information gleaned from this process. The purpose of the process is to help students learn more efficiently; if the process doesn't facilitate focused, intentional learning, then it isn't doing its job. Also, an overarching goal of formative assessment is to help students learn how to regulate their learning themselves, by self-monitoring their understanding and procedural fluency and seeking ways to improve in those places where improvement is needed.

We found that many teachers began their work with us with an image of formative assessment as a useful diagnostic tool *they* could use to evaluate students' understanding. They saw themselves as the keepers of and consumers of the information gathered from formative assessment practices. Their experience with formative assessment centered primarily on trying to figure out "Is students' understanding accurate?" (determining *whether* students understand what was intended) and, if not, then wondering "What is the nature of the problem with their understanding?" (determining both *what* it is that students do or don't understand and *why*, so that they could determine appropriate next steps for the students).

However, the heart and soul of effective implementation of formative assessment lies in the answer to the next question, "What do we do to advance the student's learning?" What is most critical in this question is the *we*, because it refers to the teacher *and to the student*. If the teacher is the only keeper and consumer of this information, then that question becomes merely "What do I, the teacher, need to do about this?" The most important person in the interaction, the student, is already in danger of being left behind in spite of the best intentions of any teacher, because while the teacher can adjust instruction, it is the student who has to adjust his or her learning. Thus, an important goal of the use of formative assessment is to provide students an opportunity and a means to learn and internalize self-regulation skills and thereby take ownership of their learning.

Comparing Formative Assessment Practices to Your Teaching Practices

When you look at our description of formative assessment, you undoubtedly recognized much of it as something you already do in your classroom:

You get information from your students, interpret that information, make decisions based on it, and provide feedback. You might even have a standard practice of launching your lessons by telling your students what you expect them to learn. How, then, is this different from your good teaching practices?

Although the differences may seem subtle, formative assessment provides additional focus and purpose to these typical teaching practices that can make a big difference for you and for your students. Consider Table 1.1, which illuminates these differences. Chapters 2 through 7 will help you understand what these differences mean and how to incorporate them into your own classroom practices.

Table 1.1 Differences Between Typical Practices and Formative Assessment Practices

What teachers do as part of instruction:	What teachers using formative assessment practices add to instruction:
Post lesson learning targets for students.	Help ensure that students understand what the learning intention itself means.
	Provide success criteria that specify what it means to meet the learning intention.
	Develop a very clear picture of what it will look like or sound like when a student meets any one of the success criteria.
	Over the course of the lesson, share that information with students so that they too know what it looks like and sounds like to meet each success criteria.
Assess students (get information about how they're doing).	Focus the assessment information gathered toward the learning intention through the use of the success criteria.
	Gather information about students' development of conceptual understanding as well as of mastery of procedures.
Interpret assessment data and make decisions based on that data.	Use the success criteria as a framework or guide for interpretation of data.
	Use a learning progression for the math topic to inform instructional decisions about where students need to go next in their learning.
	Use a learning progression to inform teacher's thinking about the nature of students' difficulties.
Provide feedback to students on their progress.	Provide a type of feedback that helps the student understand (a) how his or her work or understanding of a concept does or does not meet the learning intention by referencing the success criteria and (b) what to do next if he or she has not yet met the learning intention.
	Provide opportunities for students to act on the feedback.
Encourage students to take responsibility for their learning.	Use specific instructional techniques to help students learn how to take responsibility for their learning.
	Give students repeated opportunities to practice and develop new habits and skills that help students become more self-regulating.

■ TEACHING STUDENTS HOW TO PARTICIPATE IN FORMATIVE ASSESSMENT

Students can gain ownership of their learning and become self-regulating learners as part of your implementation of formative assessment, and throughout the rest of this book, we talk about this idea in depth. However, there are some important points to raise here that you will hear echoed in later chapters:

- *Students need guidance about what it is they are expected to do for each part of formative assessment.*

  As we worked with teachers, it became clear that helping them hone their use of various formative assessment practices was not enough; students were not yet engaging as fully as we wanted. We learned that students needed to be taught certain skills and mindsets to help them participate effectively in the cycle. Throughout this book, we've shared a collection of the tools we developed and used to help students do so.

- *Students benefit from teachers being explicit with them about the purpose of various parts of formative assessment.*

  The more that the teachers we worked with were able to talk with students about what they were doing together, and why, the more students began to buy in and participate. Explicitly sharing the purposes of formative assessment practices turned out to be one important way to help students also become consumers of the information of formative assessment.

> ### TEACHERS' VOICES
>
> I realized that when I became more transparent to my students about my actions—helping them see what I was doing during a lesson and why—they were able to get involved. When I did a short lesson on the concept of formative feedback—what it is and how we will use it in class—and then explicitly modeled it, students totally caught on. The level of discourse and engagement grew. It makes so much sense that for students to do this, they need to be taught. I don't know why I didn't think of this before. After all, we have been working on learning to use formative assessment for a year now. Students need instruction about the process, too.

- *Students need time to learn to do these things.*

  Needing time includes providing some class time to introduce students to the critical aspects. But perhaps more important, it refers to allowing students an adequate amount of time to build the understanding of their role over time. Teachers in our professional development program spread this learning out over several months, introducing one piece at a time and giving students opportunities to practice each piece with guidance, so eventually they could internalize it.

Resources for Teaching
Students About Formative Assessment

In cooperation with some of the teachers we worked with, we have developed some lessons to introduce students to learning intentions and success criteria and to formative feedback. In the corresponding subsequent chapters, you'll have the opportunity to look at these lessons and consider using them.

Introducing students to the elicitation and interpretation of evidence is a little subtler. Rather than a lesson on evidence, each individual strategy for eliciting evidence (such as the "X-Marks-the-Spot" strategy in which students self-assess using a particular template or the "Reflect Aloud" strategy in which the teacher models a way for students to share their mathematical thinking) needs to be explicitly introduced; sharing the purpose of those activities with your students will help them understand what you're asking for and why honest responses are important, and it will help them see how they may also be able to start to assess themselves. For example, students are familiar with teachers gathering evidence through questioning and quizzes, although they may assume that follow-up questions from a teacher during class mean they have provided incorrect answers. In time, they can overcome this assumption, if you talk with your students early and often about your desire to understand their thinking, whether they provide correct or incorrect answers, *and* follow up by asking about that thinking! Over time, as they gain trust that you truly are interested in their thinking, they will become more confident in and more practiced at explaining their thinking.

Once students have been introduced to these aspects of formative assessment, another important thing you can do throughout your instruction is to model for students the parts of the process that they need to learn to do. For example, you can refer to the success criteria when you elicit evidence so students understand what you're trying to assess; then, you can refer to it again when you tell them what you have discovered about the extent to which they meet the criteria (i.e., your interpretation of the evidence). As students see how you use these things to help them learn, they can begin to internalize their use.

Finally, with those understandings, students can begin to learn how to self-assess and make decisions for themselves about what they need to do to progress in their learning. You will need to help them learn to self-assess their own progress *based on evidence*, using the learning intentions and success criteria, and we discuss this in greater depth in Chapter 5, "Developing Student Ownership and Involvement in Your Students." Determining their best option for moving forward then requires identifying, and having access to, a variety of resources—their textbook, their notes and past work, the library or the Internet, their peers, and their teacher. You can help them build a repertoire of resources by varying your own suggestions for choosing which resources to move them forward and gradually turning over to them the responsibility for choosing appropriate resources.

■ HOW TO USE THIS BOOK TO LEARN WHAT *YOU* WANT TO LEARN

You may find that one or more aspects of formative assessment resonate with you or interest you as a place to begin to learn more. You may be someone who prefers to work from start to finish in order, reading the chapters in chronological order. Or you may be someone who prefers to skip around, looking for parts of the book that address questions you have or that grab your attention. So we'd like to offer a few words of advice on how to make the most of this book:

- Read Chapters 1 and 2 first. Chapter 1 (which you may have now read already!) provides an important introduction to the comprehensive view of formative assessment on which we base this book and the importance of engaging students fully in the formative assessment process. Chapter 2 discusses the nature of the importance of establishing learning intentions and success criteria as well as the importance of establishing them for use *with students*. Because the learning intention and success criteria are foundational to implementing the other aspects of formative assessment, you will be better prepared to skip around to various chapters after you have read Chapter 2.

- There is a method to the order in which the chapters are presented that may inform your choices about how to read this book:

  o Chapters 2, 3, and 4 describe an important triad of critical aspects: writing and sharing a learning intention and success criteria, eliciting and interpreting evidence and choosing an appropriate responsive action, and articulating and using formative feedback. These three elements together form the backbone of any effective implementation of formative assessment. There is more weight given throughout these chapters to the teacher's role, because in our work with teachers, we found that they needed to understand their part in these three aspects of formative assessment before they could fully delve into the ways to bring their students into the cycle. Therefore, you'll see some brief mention of the students' role in these chapters, but an in-depth discussion of the student role is saved for Chapter 5.

  o Chapter 5 talks further about developing student ownership and involvement in the formative assessment cycle. In this chapter, we flesh out what the students' role entails and how to help students learn to step into that role.

  o Chapters 6 and 7 go into more depth about the supporting aspects: using learning progressions and establishing a classroom environment conducive to formative assessment. We could have chosen to talk about those earlier, since clearly classroom environment and a deep understanding of one's subject matter are fundamental to any successful instruction. However, we know that teachers already are experienced with establishing classroom environments and thinking about their subject matter content, so we want to provide a clear picture of how these two aspects are important to implementing formative assessment. To do that, we need to establish a full vision of formative assessment and your students' role in it. We believe postponing discussion of these aspects until after establishing the context of formative assessment makes the discussion more relevant for you, our reader.

o Chapter 8 then steps back to discuss ways to implement this work. We suspect—and hope!—that you will want to try things during your read of various chapters, so trying some things out need not wait until Chapter 8! However, we also know that the teachers we worked with needed different kinds of opportunities to try things. They first needed opportunities to try individual ideas or strategies from each aspect of formative assessment, and we encourage you to do this as you move through the chapters. As they became more comfortable with individual aspects of formative assessment, they then needed opportunities to weave the pieces together into a more seamless whole. For some, this weaving happened as they added on each new piece; for others, it was a process of focusing on one aspect, then dropping it as they focused on a new one, then picking it up again later. This learning about weaving the pieces together is a very important part of making your implementation successful, so Chapter 8 focuses on some broader recommendations about how to build and sustain your journey toward full implementation.

However you choose to make the most use of this book, we hope that you find many useful strategies, tips, resources, and guidance for bringing your students into the formative assessment equation.

CONCLUSION ■

The following quote from Dylan Wiliam describes both the impact and the challenge of implementing formative assessment practices:

> *To be effective, [formative assessment practices] must be embedded into the day-to-day life of the classroom, and must be integrated into whatever curriculum scheme is being used. That is why there can be no recipe that will work for everyone. Each teacher will have to find a way of incorporating these ideas into their own practice, and effective formative assessment will look very different in different classrooms. It will, however, have some distinguishing features. Students will be thinking more often than they are trying to remember something; they will believe that by working hard, they get cleverer; they will understand what they are working towards; and will know how they are progressing. (2000)*

Formative assessment provides a structure for focusing your instruction, and your students' attention, on the concepts and skills you decide are most important for the lesson. Your role is to bring all the pieces into play both in planning and implementation, and then to teach students about their role—initially to engage in it, and eventually to internalize it and move toward being more self-regulating in their mathematics learning.

Our intent in this book is to illustrate the strengths of formative assessment by illuminating what it means to embed these practices into the day-to-day life of the classroom and to help you address the challenge of implementing the formative assessment cycle by providing you with some resources and tools to do so. We hope that you find it as rewarding as the teachers in our professional development program did.

■ RESOURCES

The following sections present some resources that can help you learn about formative assessment. Each of these resources is referenced in earlier sections of the chapter, but here we provide a consolidated list. All resources can be found at Resources.Corwin.com/CreightonMathFormative Assessment.

Learning Resources

This resource supports your learning about the critical and supporting aspects of formative assessment:

- **Formative Assessment Overview** provides an approach to considering different characteristics of formative assessment. The interactive web page provides several statements about formative assessment, by selected notable authors, which you can sort to look for connections among the ideas.

Reference Resources

These resources summarize key ideas about formative assessment practices:

- **Chapter 1 Summary Card** is an index-size Summary Card that provides a summary of the critical and supporting aspects of formative assessment.
- **Formative Assessment Cycle** is a color version of the *full* Formative Assessment Cycle that we are building throughout this book. While the cycle evolves over the course of the book, some of you may want to see the completed version early. We recommend using this color version, possibly with the video available on the companion site.
- **Formative Assessment Cycle Video** is an audiovisual tour of the Formative Assessment Cycle that we are building throughout this book. You may want to look at it now or wait until Chapter 8.

2

Using Mathematics Learning Intentions and Success Criteria

During her professional development work in formative assessment, Mrs. Daniels, a sixth-grade mathematics teacher, shared the following thinking in her online blog:

> I used to think that the way you were supposed to construct a good lesson was to unveil the learning for students over the course of a lesson, then at the end, sum it up, saying, "So what we've learned today is ____." Of course, all along there, my students had to trust me and just follow along where I was trying to lead them—and sometimes they wouldn't always follow, or were trying to follow and got off course and didn't know it. And sometimes I didn't know it either, certainly not right away. But I'm now realizing that by sharing the learning intention and success criteria with the students at the start of our lesson, I'm changing how things work in math class. I'm saying "Here's where we're going to try to go today; come with me, I'll help you see how to get there. And here are some signposts along the way—we're going to look for those signposts to see if we're heading in the right direction." I'm finding my students are being much more willing to "take the journey" with me when I give them some sense of where we're going today and how to tell if

they're on track. It's about starting the journey with the destination in mind, and helping them also see that destination from the start.

Mrs. Daniels makes a great point about the importance of setting, and sharing, clear learning goals for her students and clear indicators that learning is moving in the right direction. In this chapter, we look closely at what we call *learning intentions* and *success criteria*.

■ WHAT ARE LEARNING INTENTIONS AND SUCCESS CRITERIA?

Learning intention is the name used in much of the formative assessment literature for the learning goal for a lesson. It is what teachers intend for the students to learn. You may have a different name that you prefer to use with your students; the importance of learning intentions lies not in what you call them but in how you articulate them and then use them during your instruction. There are important characteristics of both the content of learning intentions and their use that may distinguish them from other types and uses of lesson goals or lesson targets, and we illustrate those characteristics in this chapter.

A brief collection of accompanying statements support the learning intention by giving both the students and their teacher guidance on what to look for in students' learning if that learning is on track toward reaching the learning intention. These *success criteria* describe characteristics of what students will say, do, or produce and are tangible and therefore observable. However, that does not mean they are necessarily procedural! (We illustrate what this means as well throughout this chapter.)

In Chapter 1, "Using Formative Assessment to Build Student Engagement in Mathematics Learning," we discussed that formative assessment can be a vehicle to help students to become self-regulating learners. We introduced the "Thinking Like a Self-Regulating Learner" diagram (see Figure 2.1) to summarize the key questions and actions that

Figure 2.1   Learning Intentions and Success Criteria in Self-Regulation

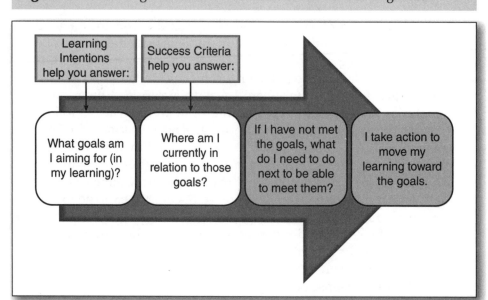

self-regulating learners internalize and use to guide their learning. Mrs. Daniels talked about using learning intentions to provide the first step in that process, a clear understanding of the learning intention the students should aim for. With corresponding success criteria to guide their self-assessment, students can also begin answering the question, "Where am I currently in relation to those goals?"

In Figure 2.2, we are expanding our Formative Assessment Cycle to include more detail about how a learning intention and success criteria appear in a classroom lesson.

Notice that we are adding banners across the top to define three sections: before the lesson, during the lesson, and after the lesson. Determining the learning intention and success criteria appears both at the beginning of the Formative Assessment Cycle, before the lesson, and at the end, after the lesson—the end of one cycle is effectively the beginning of the next. During the lesson, the teacher shares the learning intention and success criteria with the students. The students' role is to understand them; providing time for students to process and ask for clarification is always helpful.

Example: Recall Ms. Jenkins's lesson on rates of change from Chapter 1. She wants students to work toward recognizing how constant change can look in different representations. Before the lesson, in her planning, she decides on this learning intention: "Today, we will learn to relate growth in numeric patterns to growth in corresponding visual patterns." For success criteria, she decides on the following:

1. *I can describe how a pattern of numbers continues.*

2. *I can describe how a corresponding visual pattern continues.*

3. *I can explain how the way a number pattern grows relates to the way the visual pattern grows.*

In the classroom, the lesson starts with the warm-up activity. After her students have completed the activity, Ms. Jenkins reminds them of the problems they worked on the previous day, creating tables of values from a growing visual pattern. She introduces the learning intention and explains how this is similar to, and different from, the previous work. She gets volunteers to read the success criteria, and then she directs the students to turn to a partner and discuss how success with each criterion shows progress toward the learning intention. After a few minutes, during which Ms. Jenkins listens to several of the discussions, she calls on a few students to share what they talked about. Then she gives them a chance to ask any questions they may have about the learning intention or the criteria.

Throughout the lesson and at its conclusion, Ms. Jenkins and her students will use the success criteria as they assess the students' progress in learning and act to help the students reach the learning intention. Ms. Jenkins will use the experiences of the day's lesson to plan for instruction that follows, including revising the learning intention and success criteria—or creating new ones—based on students' progress in learning.

As you learn more about the other critical aspects of formative assessment, you will see how developing solid learning intentions and success

26

diagram will lend itself to
TKU guide

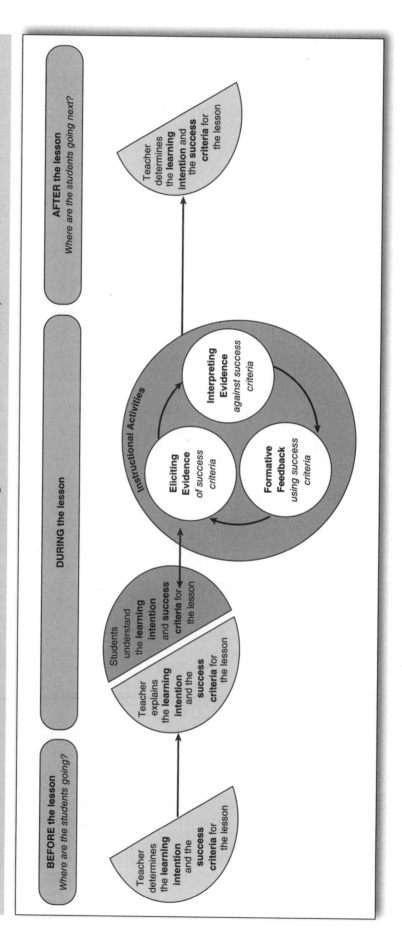

Figure 2.2 The Evolving Formative Assessment Cycle (with learning intentions and success criteria)

BEFORE the lesson
Where are the students going?

Teacher determines the learning intention and the **success criteria** for the lesson

DURING the lesson

Teacher explains the **learning intention** and the **success criteria** for the lesson

Students understand the **learning intention** and **success criteria** for the lesson

Instructional Activities

Eliciting Evidence *of success criteria*

Interpreting Evidence *against success criteria*

Formative Feedback *using success criteria*

AFTER the lesson
Where are the students going next?

Teacher determines the **learning intention** and the **success criteria** for the lesson criteria, and sharing them with students, forms the backbone of all the formative assessment practices.

Characteristics of Learning Intentions and Success Criteria

Many teachers share the goal for a lesson with their students by posting daily objectives, SMART goals, measurable objectives, mastery objectives, learning targets, or "I can . . . " statements. Here are some examples of lesson goals you might see in a middle school math classroom:

- *Today's Target: I can create a histogram for a set of data.*
- *Students will be able to: Substitute for a variable and find a value for an expression*
- *Today we are working toward Standard 6.RP.3: Use ratio and rate reasoning to solve real-world and mathematical problems, for example, by reasoning about tables of equivalent ratios, tape diagrams, double number line diagrams, or equations*

In some cases, the learning goal may be implied, though not explicitly stated, in the day's agenda:

- *Today's Agenda*
  - *Do Now 3*
  - *Review equivalent fractions using fraction strips*
  - *Use fraction strips to add fractions with unlike denominators (pages 34–35)*
  - *Practice problems 2–7, 9, 12–15*

Collectively, these examples reflect common characteristics of a lesson goal: measurable, understandable by students, aligned to standards, or clearly related to the lesson's activities. Each example points to successful mastery of a mathematics skill, process, or procedure—but they are not learning intentions. Each one lacks one or more of the important characteristics of learning intentions that are essential for their use in formative assessment. This section identifies the characteristics of learning intentions and their associated success criteria.

Learning Intentions Focus on the Learning, Not the Activities

The first characteristic is the most notable, and it deserves special attention:

1. ***The learning intention articulates the key understanding that a student will gain in a lesson, rather than the particular activity to be completed.***
   Let's take a closer look at a couple of the preceding examples. They do a good job of articulating the lesson *activities*—what students will *do* during the lesson—but they are not explicit about what students will come to *understand* mathematically as a result of all that doing.

Example 1: Ms. Greenberg posts a learning target for her sixth-grade math class today:

- *I can create a histogram for a set of data.*

This points very clearly to what Ms. Greenberg wants her students to be able to *do,* but it is not a clear statement about what she wants them to *understand* as a result of having mastered those skills. If we pose the questions, *"Why* do you want your students to be able to create a histogram? What important mathematical ideas do you want them to understand as a result of learning to do this?" she may say, "Because I want them to understand that data can be summarized using different representations of the data, and this is one of those representations." Or she may say, "Because I want them to understand what kinds of data lend themselves more readily to use of a histogram, rather than a different representation." Or she may say, "Because this skill is an important one to have as part of being mathematically literate." All three reasons are good ones, but it is her first two responses that reflect what she wants her students to *understand* as a result of creating histograms, and either could form the basis of her learning intention. She might then use her "I can . . ." statement, or something similar, to specify part of her success criteria for reaching her learning intention.

Example 2: Mr. Petrelli identifies a mastery objective for his algebra lesson:

- *Students will be able to substitute for a variable and find the value of an expression.*

This is a clear statement of a mathematical skill he wants his students to master successfully. If we asked Mr. Petrelli what he wants his students to understand about variables and expressions from being able to do this skill, he might say, "I want them to understand that they can substitute any number into an expression, and that the expression will have a different value depending on the number you substitute into it." Or he might say, "I want them to understand that expressions are one of several ways in which variables are used, and that the variable can take on multiple values when used in that way." Both of these answers articulate the underlying conceptual understanding that Mr. Petrelli might be seeking in addition to the mastery of the mechanics of evaluating expressions for a particular value. This conceptual understanding would become the substance of his learning intention, and his statement about what students will be able to do might be part of his set of success criteria for reaching the learning intention:

LEARNING INTENTION AND SUCCESS CRITERIA

Learning Intention: Algebraic expressions can be evaluated using different values for the variable(s) in the expression.

Success Criteria:

1. I can explain why an expression might have different values.

2. I can substitute different numbers into an expression to give it different values.

In both cases, if the teachers were able to describe to their students both the skills they need to be able to perform as well as the deeper understandings that underlie those skills, their students might be better positioned to develop those understandings. If we are to be successful at communicating these kinds of mathematics goals to students, we need to balance identifying and articulating the more intangible learning underlying a mathematical idea with describing tangible, observable instances of what it would look like and sound like when students are progressing successfully toward that learning. This balance is what learning intentions and success criteria are intended to achieve.

The goal of achieving this balance is not a new one; it's been said in many ways over the past decades. The National Research Council, in the book *Adding It Up*, said,

> *Helping all students learn to think mathematically is a new and ambitious goal, but the circumstances of modern life demand that society embrace it. . . . The research over the past two decades, much of which is synthesized in this report, convinces us that all students can learn to think mathematically. (Kilpatrick, Swafford, & Findell, 2001, p. 16)*

Many state frameworks include an emphasis on attending to building strong conceptual understanding, and the Common Core State Standards (National Governors Association Center for Best Practices [NGACBP] & Council of Chief State School Officers [CCSSO], 2010) have articulated conceptual understandings explicitly in their various content standards, some of which are listed below:

> *6.RP.2: 2. Understand the concept of a unit rate* **a/b** *associated with a ratio* **a:b** *with* **b** *≠ 0, and use rate language in the context of a ratio relationship.*

> *6.NS. 7: Interpret statements of inequality as statements about the relative position of two numbers on a number line diagram.*

> *7.EE.2: Understand that rewriting an expression in different forms in a problem context can shed light on the problem and how the quantities in it are related.*

> *7.SP.1: Understand that statistics can be used to gain information about a population by examining a sample of the population; generalizations about a population from a sample are valid only if the sample is representative of that population. Understand that random sampling tends to produce representative samples and support valid inferences.*

> *8.F.1: Understand that a function is a rule that assigns to each input exactly one output. The graph of a function is the set of ordered pairs consisting of an input and the corresponding output.*

> *8.G.5: Use informal arguments to establish facts about the angle sum and exterior angle of triangles, about the angles created when parallel lines are cut by a transversal, and the angle-angle criterion for similarity of triangles. For example, arrange three copies of the same triangle so that the sum of the three angles appears to form a line, and give an argument in terms of transversals why this is so.*

Each of these standards points to a sophisticated mathematical understanding that goes beyond being able to complete certain kinds of problems accurately. The Standards for Mathematical Practice, provided as part of the Common Core State Standards, outline a powerful set of habits students can learn to use in order to develop stronger mathematical reasoning abilities.

COMMON CORE CONNECTION

The term *understand* can seem vague or problematic to some, but authors of the Common Core Standards underscore an important message that we echo in this chapter:

These standards define what students should understand and be able to do in their study of mathematics. But asking a student to understand something also means asking a teacher to assess whether the student has understood it. But what does mathematical understanding look like? One way for teachers to do that is to ask the student to justify, in a way that is appropriate to the student's mathematical maturity, why a particular mathematical statement is true or where a mathematical rule comes from. Mathematical understanding and procedural skill are equally important, and both are assessable using mathematical tasks of sufficient richness. (NGACBP & CCSSO, 2010, p. 4)

Striking a balance between developing skills and conceptual understanding sometimes can be difficult to do in a mathematics classroom. We have met and worked with many teachers who have a strong commitment to developing a balance between procedural fluency and strong conceptual understanding in their students, so we want to underscore that there is no lack of this commitment on the part of teachers. However, providing *explicit statements to students* of the learning goals related to *conceptual* understanding of mathematics, in ways that students can understand and therefore try to learn, was much more difficult for teachers than being explicit with students about the skills and procedures they needed to master. While some may believe that students will just develop understanding of these underlying conceptual ideas from doing the related problems, we do not share that belief. Based on our years in our own classrooms and in the classrooms of other teachers with whom we've worked, we have seen ample evidence that an overabundance of attention to developing skills does not build a classroom of strong mathematical thinkers who have a robust understanding of mathematics.

As a result, we have come to believe that mathematical learning intentions can, and should, remain focused on deeper conceptual understanding because they serve as a way to make explicit for students what the important conceptual learning is. This does not mean that we value the learning of mathematics procedures any less; on the contrary, we believe that procedural fluency is vital to being able to use mathematics flexibly in a variety of problem situations. Rather, it means that we believe that it is essential to be explicit with students about what pieces of conceptual understanding we want them to learn, as well as being explicit about what acquisition of that understanding will look like and sound like. And therein lies the real challenge in creating and using learning intentions and success criteria effectively.

Additional Key Characteristics of Learning Intentions and Success Criteria

The other key characteristics that distinguish learning intentions and success criteria from other types of goals and objectives are as follows:

2. The learning intention focuses the lesson on the highest-priority learning for that lesson.

3. The success criteria describe examples of something a student will be able to say, do, or produce if the student's learning is on track toward reaching the learning intention. They are tangible or observable. (However, not everything that is tangible or observable is a success criterion.)

4. The success criteria, collectively, provide enough evidence to make both the teacher and students reasonably confident that students have reached the learning intention.

5. The learning intention and the success criteria are aligned to each other.

6. Both the learning intention and success criteria are written to be understandable by students.

Let's look at each of these characteristics more closely.

2. *The learning intention focuses the lesson on the highest-priority learning for that lesson.*

With the learning intention, you are saying to students, "Of all the things we might think about or discuss today, this is the understanding that I most want you to try to develop today. It gets priority." By prioritizing the point of a lesson during your planning and then articulating it clearly during the lesson, you create a focus that is useful to both you *and* your students.

For example, suppose Mr. Petrelli's algebra lesson (Example 2 earlier) is part of an introductory algebra unit or chapter that includes learning about some of the different uses of variables and what a variable means. In this particular lesson within the unit, he thinks about the question, "What is the main idea or understanding that I want my students to take away from today's lesson?" and decides to focus the learning intention on one of several uses of a variable: the idea that variables can take on multiple values when used in an algebraic expression. During the lesson, the class may well talk about or do many things related to variables, such as what it means to *substitute*, the difference between an expression and an equation, what are other examples of variables students have seen (such as in area formulas), or even what it means to *find* x or *find* n. They may also spend time practicing how to substitute different numbers for the variable and finding the value of an expression, recalling order of operations, and reviewing what an exponent means. For some students, particularly students struggling with a math concept, it can be difficult to figure out what exactly they're supposed to be learning. In the absence of understanding how to prioritize that long list of skills and ideas, the ideas can quickly become a disparate collection of different things that all seem equally important (or unimportant). Lacking a point of focus that would help the students more clearly understand where their energies and concentration should be placed, some may just give up and disengage—it's just too much for them. Articulating a learning intention about a key idea that gets priority in the lesson provides that focus for students. They may well learn some of the other ideas that arise during the lesson, but the focus of the learning intention helps students understand where to focus their efforts and their thinking.

It can also provide focus for you during instruction. In Mr. Petrelli's case, it could be easy to get lost in trying to juggle the needs of students who needed to review order of operations, who didn't remember what an exponent is, or who didn't understand how to substitute correctly. This can lead Mr. Petrelli to be consumed with helping students successfully complete a collection of procedures that they struggled with. Having the learning intention clearly stated—and *using it in a variety of ways* to ground the learning (which we discuss in Chapter 5, "Developing Student Ownership and Involvement in Your Students")—can help you keep the learning focused on maintaining that balance between developing procedural fluency as well as conceptual understanding.

In our work, we found that such a focus was best achieved when the teacher kept to one learning intention for a lesson or for a defined set of instructional activities. For this reason and for consistency, the examples we provide throughout this book include only one learning intention for a lesson. This is not a hard-and-fast rule of formative assessment, though, and teachers sometimes find it necessary to have more than one learning intention.

3. *The success criteria describe examples of something a student will be able to say, do, or produce if the student's learning is on track toward reaching the learning intention. They are tangible or observable. (However, that does not mean that everything that is tangible or observable is a success criterion.)*

The role of the success criteria is both (1) to help students understand what to pay attention to in their own learning during the lesson so that they may gauge how well they are reaching the learning intention and (2) to help you add further focus to the lesson by articulating what the possible evidence would be of reaching that learning intention. In coming up with the success criteria, a teacher needs to think carefully about the question, "If my students reached the learning intention, what would I see them do, and what would I hear them say?"

An important caveat here, however, is that while success criteria are tangible or observable, not all tangible or observable things are necessarily success criteria. For example, in Mr. Petrelli's algebra lesson, he could certainly observe whether students could perform order of operations correctly or provide an accurate definition of an exponent. However, neither of those skills is necessarily an indication that his students were working toward reaching the learning intention of understanding that variables can take on multiple values when used in an algebraic expression. While they may be useful or even necessary skills to bring to bear, that does not necessarily make them success criteria. Instead, when Mr. Petrelli thinks about what he would see his students say, do, or produce that would feel to him like indications of reaching the learning intention, he decides that he would look for students being able to explain why an expression might have different values and to substitute different numbers into an expression to give it different values. These would become his success criteria for this lesson.

4. *The success criteria, collectively, provide enough of an indication to make both the teacher and students reasonably confident that students have reached the learning intention.*

In Mr. Petrelli's example, you may be wondering why his list of success criteria seems short or why he hasn't included other possible success criteria. Once you have articulated the learning intention for a lesson, you have to ask yourself, "What would be *enough* evidence that students had reached the learning intention, without being an overwhelming list?" Part of this answer has to do with making sure that the success criteria point to both mastery of any relevant procedures and conceptual understanding of the main idea of the lesson. It's easy to create success criteria that point to mastery of procedures; many teachers already use mastery objectives or "I can . . ." statements in this way. However, since the production of correct answers is not necessarily an indication that understanding is developing, additional success criteria are needed. For example, suppose you want students to understand why dividing the numerator and denominator by a common factor produces an equivalent fraction (conceptual understanding). You might ask students to find the simplest form for various fractions, such as $\frac{16}{64}$ or $\frac{15}{25}$. Many students will be able to do this correctly by following a procedure (demonstrating procedural fluency), but this alone does not show that they know *why* the procedure works.

Therefore, it's also important to ask students to explain, describe, justify, or show something about their understanding. The first consideration for whether the success criteria provide enough indication is that they include pointers to conceptual understanding as well as to procedural fluency, and they do not point just to procedural fluency. We refer to these as *procedural success criteria,* versus *conceptual success criteria.*

The other consideration for whether the success criteria provide enough indication relates to their purpose. The success criteria describe for both you and your students what you and they will look for as indications of reaching the learning intention. Students need enough opportunity to produce evidence of each success criterion during the lesson. If you include too many success criteria, you risk creating a list too long to address in a single lesson and defeat the purpose of giving the lesson focus. Generally, we have found that two or three success criteria are often manageable during a lesson. We say more about this in Chapter 3, "Gathering, Interpreting, and Acting on Evidence," as we talk about using the success criteria to guide your collection and interpretation of evidence of students' learning.

In thinking about these considerations, it's important to note that reaching a learning intention means meeting *all* of the corresponding success criteria; taken as a set collectively, meeting the success criteria points to reaching the learning intention. In a given class period, for example, most of your students may have met (or even partially met) one success criterion but still need more time before they can meet the others. That is not uncommon and very reasonable to expect, given that learning happens at different rates and in different ways for your students.

5. *The learning intention and the success criteria are aligned to each other.*

Part of determining your learning intention and success criteria involves posing to yourself a question such as "What are the *right* criteria to show that students have reached the learning intention?" If you include success criteria that have at best a peripheral connection to the learning intention, then meeting the success criteria won't necessarily lead to students reaching the learning intention (at least, not the one you were aiming for!). Let's compare a set of success criteria that do align with a learning intention to some criteria that do not. Mr. Petrelli's learning intention and success criteria that do align with each other are:

LEARNING INTENTION AND SUCCESS CRITERIA

Learning Intention: Variables can take on multiple values when used in an algebraic expression.

Success Criteria:

1. I can explain why an expression might have different values.

2. I can substitute different numbers into an expression to give it different values.

He could have come up with the following success criteria that, while they address related ideas, would not align to his learning intention:

Success Criteria That Do Not Align	Why Not?
• I can solve for a variable in a simple equation.	This lesson is focused on the use of variables in expressions, where they can take on multiple values. This criterion focuses on a different use of variables (as a single, unknown value).
• I can create an algebraic expression to describe a situation.	While this is an important skill related to algebraic expressions, the focus of this lesson is on variables taking on multiple values in an expression. This idea distracts from that central focus.
• I can explain other uses of variables.	This might be an appropriate criterion for a later lesson when other uses of variables have been explored, but this lesson focuses only on how variables are used in expressions.

None of these success criteria is bad—they all describe indicators of important learning related to variables and expressions. Rather, the main point here is that success criteria are most effective when they *help maintain the central focus* suggested by the learning intention. While the initial question for this characteristic, "What are the *right* criteria to show that students have reached the learning intention?" is a very natural and common one for you to ask yourself, we think an even more helpful question would be "What success criteria maintain the same focus as my learning intention?"

6. *Both the learning intention and success criteria are written to be understandable by students.*

Because your students are the primary audience for your learning intention and success criteria, both need to be written so that students can understand them. In some circumstances, you may want to downplay the use of formal mathematical language, depending on your students and your goal for the lesson. However, understandable by students does not necessarily mean nonmathematical. In many circumstances, you may want to highlight certain mathematical or academic language that may be unfamiliar to students, especially when part of the goal of the lesson is to develop understanding and use of that language. For example, a learning intention that reads *"Today, I will learn what the slope of a line means and what it has to do with rate of change"* includes mathematical language that students will learn about in the lesson (or may have been introduced to in the previous lesson) and, when shared and discussed, can be understandable while still retaining useful mathematical language. The questions to consider for this characteristic are, "Are my learning intention and success criteria written in a way that my students can understand? If new terminology is included, how will they learn about that new terminology?"

These characteristics provide guidelines for distinguishing learning intentions and success criteria from other types of goals. We have created additional examples of learning intentions and success criteria for you to consider, which are available on the companion resource website (see below). The characteristics also provide guidelines for creating your own, and in the next section, we delve into a more specific discussion of how you can create them yourself and provide a first pass at a discussion of how to use them. (We take up the rest of that discussion in Chapter 5, "Developing Student Ownership and Involvement in Your Students.")

As we noted near the beginning of this chapter, you do not need to call them *learning intention* and *success criteria*. Many of the teachers we worked with used this language with their students, but some did not. How you label and present the learning intention and success criteria for a lesson needs to work for you and your students, so that your students can understand their purpose and use them. Throughout this book, we use a variety of formats to demonstrate that communicating the learning intention and success criteria to your students is more important than the labels you use to express them.

 Learning Resource: Go to Resources.Corwin.com/
CreightonMathFormativeAssessment to access this resource.

 Examples of Learning Intentions and Success Criteria. This sortable database provides additional examples of learning intentions and associated success criteria.

■ CREATING LEARNING INTENTIONS AND SUCCESS CRITERIA FOR YOUR CLASSROOM

As you may have noticed in the previous section, a question was included with each characteristic that can help you plan your own learning intentions and success criteria. Table 2.1 summarizes these questions alongside their corresponding characteristics.

Writing learning intentions and success criteria that meet all these characteristics can feel daunting at first. Drawing on the experiences of the teachers in our professional development program, we can assure you of two things:

1. These characteristics are intended to lay out for you the goal to aim for with an effective learning intention and accompanying success criteria, but they describe the endpoint, not your own starting point. You will improve with time and practice.

2. No students will be harmed if your learning intentions and success criteria are not perfect. Don't be deterred from putting these out

there for fear that they are not perfect. It is through your experiences in using them and seeing how students interact with them that you will get better.

Table 2.1 Summary of Characteristics and Related Questions

Characteristic	Questions to Ask Yourself
The learning intention articulates the key understanding that a student will gain in a lesson, rather than the particular activity to be completed.	What is it that I want students to understand about the problems they're solving?
The learning intention focuses the lesson on the highest-priority learning for that lesson	What is the main idea or understanding that I want my students to take away from today's lesson?
The success criteria describe examples of something a student will be able to say, do, or produce if the student's learning is on track toward reaching the learning intention. They are tangible or observable.	If my students reached the learning intention, what would I see them do and what would I hear them say?
The success criteria, collectively, are sufficient to make both the teacher and students confident that students have achieved the learning intention.	What would be _enough_ criteria that students had reached the learning intention, without being an overwhelming list?
The learning intention and the success criteria are aligned to each other.	How can I articulate success criteria that maintain the same focus as my learning intention?
Both the learning intention and success criteria are written to be understandable by students.	Are my learning intention and success criteria written in a way that my students can understand? If there is new terminology included, how will they learn about that new terminology?

TEACHERS' VOICES

When I first started writing learning intentions and success criteria, I felt like I was perseverating over getting them perfect. I wanted it to be A+. Once I started using them in the classroom, I realized that they didn't have to be exactly perfect. I could tweak them or fix them, and even in the middle of a lesson, I've erased a certain word and replaced it with another and the students don't mind. Actually, it helps them. Don't perseverate over getting them exactly perfect. Depending upon your students, the learning intentions and success criteria are going to need to look different. You might end up changing them. It's fine. It's all good.

Your Role Before the Lesson: Planning With a Learning Intention and Success Criteria in Mind

In our Formative Assessment Cycle, there is one step that falls under the Before the Lesson banner. (See Figure 2.3.) Prior to the start of a lesson, you'll want to articulate your learning intentions and success criteria for students.

Figure 2.3   The Formative Assessment Cycle, Before the Lesson

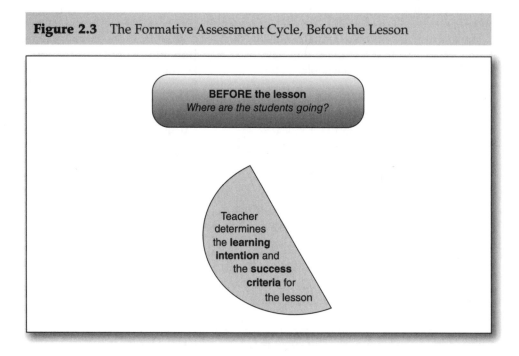

Before we present some recommendations on how to complete that important first step, we'd like to note a shift in teachers' thinking as they started creating learning intentions and success criteria and trying to use them on a regular basis. During her professional development work with formative assessment, Ms. Fallon, an eighth-grade mathematics teacher, described the shift that she was experiencing, summarized in Figure 2.4. She talked about how this shift occurred in her expectations for her students, her thinking about her lesson planning, and in how she articulated the learning objective (and later the learning intention and success criteria).

Several things are striking about the shift in Ms. Fallon's thinking represented here. For example, she now thinks of her students, not just herself, as having responsibility for knowing what they have learned. ("How will both the students and I know what they have learned?") This is absent from her "Before" way of thinking about planning. Some students undoubtedly do think about whether they are learning, and she may have expected it from some students before this shift. However, after the shift, she is being explicit that she *does* expect it, and she expects it from *all* her students.

Her planning questions reflect yet another significant shift. Her "Before" questions—*What topic will I teach? What activities will support this? What will be tomorrow's objective?*—focus on the topic and the relevant

Figure 2.4   A Shift in Thinking about Planning

BEFORE I began to use formative assessment practices	NOW that I use formative assessment practices
I asked myself these questions to guide my planning: • What topic will I teach? • What activities will support this? • What will be tomorrow's objective?	I ask myself these questions to guide my planning: • What is it that I want the students to learn? • How will both the students and I know what they have learned? • How will this direct tomorrow's instruction?
I wrote learning objectives like this: Learning Objective: *Students will be able to create tables and graphs from a given data set.*	I write my goals like this: Learning Intention: *I will understand how a table and graph represent the same data.* Success Criteria • *I can create a table using data.* • *I can create a graph using the same data.* • *I can explain how the same data are represented both in the table and on the graph.*

activities, without clear attention to the learning that she wants to occur as a result of the activities. Her "Now" planning questions—*What is it that I want the students to learn? How will both the students and I know what they have learned? How will this direct tomorrow's instruction?*—put the students' learning in the lead and her planning decisions about instruction follow from that. She certainly still needs to think about the lesson's activities during planning. However, the series of activities are not what she now uses to organize the lesson; the intended learning serves as the primary organizer of the lesson, with activities chosen to serve and support that learning.

As a result of this shift, the writing of her lesson objectives changed as well. While her previous lesson objective was clear, observable, and a relevant and valid mathematical goal for students, it was framed in terms only of what students would *do*, not what they would *learn*. Ms. Fallon assumed that if students were able to master these mathematical skills, they would also develop an understanding about the relationship between tables and graphs. As her thinking about her lesson planning shifted, her lesson objectives shifted as well to learning intentions that became more explicit about the underlying conceptual learning she wanted students to gain. She began to think in terms not of the intended activities for the lesson, such as creating tables and graphs, but of the intended learning that would result from having participated in those activities, such as how tables and graphs represent the same data. This resulted in a learning intention that stated in a clear and student-friendly way precisely what she wanted students to learn in a given lesson.

She also knew it was important to spell out some measurable indicators that progress was being made, much like her previous lesson objectives. The three success criteria listed below the learning intention spell out clearly for her, and for her students, what she and they will see if students' learning is progressing appropriately. Her first two success criteria point to skills that students would be able to demonstrate, and her third success criteria tells students they also need to be able to explain something about the key point of the learning for that lesson ("I can explain how the numbers in the table show up in the graph, and vice versa.")

As you begin working with learning intentions and success criteria, you may find your own thinking shift, but don't worry about making such a shift all at once. The best way to learn about using a learning intention and success criteria in a lesson is to start trying it out, so we encourage you to be bold about trying things out, using the recommendations in the rest of this chapter to guide you, and allow yourself a learning curve. In the end, your students only stand to benefit from your efforts.

The following sections provide some practical recommendations to help focus your planning efforts around a learning intention and success criteria, as well as resources to help you create learning intentions and success criteria. You may notice that these recommendations closely parallel the six characteristics and include additional examples, information, and resources for you to consider as you build toward writing learning intentions and success criteria and using them with your students.

The Most Important Learning

As you think about a learning intention that articulates the most important learning you want your students to gain from a lesson, there are three recommendations to consider:

> Recommendation 1: Consider the mathematical understanding you want students to gain as a result of being able to perform certain procedures.
>
> Recommendation 2: Focus on the highest-priority learning for students to attain—or to make progress toward—during the lesson.
>
> Recommendation 3: Keep your learning intention manageable. Narrow your focus to something that is attainable within a lesson or just a few lessons (understanding that a lesson may span more than one class period).

As we begin looking at these recommendations, let's consider one math class.

Mr. Royce posts his objective daily in the form of a *SWBAT*— Students Will Be Able To . . . Mr. Royce tries to frame each of his lesson objectives so that they are specific, measurable, attainable, relevant, and timely. In spite of his careful attention to frame the lesson for students, he finds himself often dissatisfied with what he describes as his students not really "owning" the material—understanding it thoroughly and deeply enough that they can apply skills in unfamiliar mathematical situations or demonstrate that they "really understand it."

During one lesson, Mr. Royce posts the objective: *SWBAT set up a proportion and solve it.* One student, Evan, correctly completes the 15 problems given for homework from the textbook. He asks Evan to work with Alison, who says she "doesn't get how to set up the problems." When he comes over to check on their progress, they are working on the following problem:

> *Doug purchased 4 balloons and paid $9.75. How much would he have to pay for 18 of the same balloons?*

He hears Evan tell Alison, "4 balloons goes on top in the first fraction, then $9.75 goes on the bottom. Then 18 balloons goes on top on the other fraction and x goes in the bottom."

Alison says, "But couldn't you also put balloons in one fraction and money in the other fraction? Like 4 balloons over 18 balloons, then $9.75 over something? For that matter, couldn't you put 18 balloons on top and 4 balloons on the bottom? Does it really matter?"

Evan shakes his head. "No, you have to put the first two numbers being compared in the same fraction. You put the numbers into the proportion in the same order: 4, $9.75, 18, and x, because you don't know the other cost for balloons."

Alison looks puzzled. As Mr. Royce talks to Alison, he realizes that she understands there are different ways the proportion could be set up and can describe several of them correctly but that she is unsure which of those ways she is expected to use in solving these problems.

> **Recommendation 1: Consider the mathematical understanding you want students to gain as a result of being able to perform certain procedures.**

A good place to start in your planning is asking yourself one of the questions Ms. Fallon suggested: "What is it that I want students to learn?" Answering this question is often not as easy as it seems like it should be; after all, you have to think every day about what you want your students to learn. However, it's often much easier to articulate the procedures you'd like your students to master than it is to articulate the important mathematical ideas that underlie those procedures. If you find yourself frequently thinking, "I want my students to learn *how to* . . ." notice whether you complete that sentence more often with statements of mathematical procedures or whether you strike a balance between statements that focus on conceptual understanding and procedural fluency.

Consider the example with Mr. Royce. Although it appears at first that Alison hasn't achieved the objective (she says she's unsure about setting up the proportion) and that Evan has, Alison can interpret the relationships between the quantities correctly and can determine how to compare them. Her understanding is conceptual—she appears to have a decent working understanding of how a proportional relationship behaves—and if the question of which proportion to use is clarified, she will probably be able to apply the concept in many situations, as Mr. Royce wants his students to do.

Evan's understanding can be described as more procedural; he understands a set of steps to follow to set up a proportion, but he lacks sufficient understanding of the idea of proportionality to give him flexibility in how he sets up his solution to the problem. If a problem is worded differently so that the first two numbers coincidentally aren't related, Evan will have difficulty solving the problem correctly. For example, consider Evan's approach when faced with this problem:

> *Early in the day, Cheryl purchased 11 cartons of ice cream for the party, then started running out, and bought 4 more at the same price. If she spent $22.76 on the 4 cartons, how much did she spend on the first 11 cartons?*

Evan's line of reasoning (assigning the numbers in the proportion according to the order they appear in the problem) would result in the incorrect proportion: $\frac{11}{4} = \frac{22.76}{n}$.

In formative assessment, the learning intention is framed in terms of the highest-priority learning for that lesson. In Mr. Royce's case, he wanted his students (1) to understand different ways that a proportion could be set up, and why and (2) to be able to solve them, both (1) a conceptual goal and (2) a procedural goal. However, his posted objective is only procedural: set up a proportion, and solve it. A learning intention focused on his conceptual main idea might sound more like this:

LEARNING INTENTION

Understand why equivalent proportions can be set up in more than one way.

Recommendation 2: Focus on the highest-priority learning for students to attain—or to make progress toward—during the lesson.

Often, a math lesson will involve using a wide variety of mathematical skills and concepts. In Mr. Royce's lesson, students might be thinking not only about proportions but also about whole number and decimal multiplication and division, or unit rates. However, these are supporting mathematical ideas to the main point of the lesson. A learning intention makes clear to students what the most important focus for the learning is in the lesson.

This is beneficial for the teacher as well. Numerous teachers that we have worked with reported that as they gained facility with using learning intention and success criteria, they referred to the learning intention to keep the instruction on track when students' questions and classroom discussion broadened beyond the day's learning goal.

TEACHERS' VOICES

Having the learning intention posted where I can see it, as well as the students, has really been helpful to me in keeping me focused. Sometimes in class, we can start discussing all kinds of things related to whatever we're learning that day. For instance, I might see that the kids have a gap in something that they need in order to understand our topic that day, and I'll spend some time reviewing whatever it is that they need to fill that gap. It's easy to get distracted, but I've found that I actually use the learning intention to think, "Well, what is that I'm trying to get across here?" It brings me back to that point of focus and helps me be clearer with the kids.

As we worked with teachers, we looked for ways to help them become stronger at clearly articulating learning intentions focused on the highest-priority learning. As we looked across a variety of lessons for different content strands (such as geometry, or number sense, or algebraic thinking), we saw some recurring categories of *purposes* behind the lessons. In some cases, the lessons were focused on defining a term or understanding what a concept meant; in other cases, the lessons were designed to connect one representation (such as a table) to another (such as a graph). We found that the learning intentions and success criteria for these purposes could often (though not always) follow a very similar structure. On the companion resource website, you will find a list of these purpose categories and some guidelines for writing learning intentions and success criteria, based on these purpose categories.

 Learning Resource: Go to Resources.Corwin.com/ CreightonMathFormativeAssessment to access this resource.

 Lesson Purposes, Practice Interactive. This interactive web page helps you learn more about and practice writing learning intentions and success criteria to match an overarching purpose.

 Planning Resource: Go to Resources.Corwin.com/ CreightonMathFormativeAssessment to access this resource.

 Lesson Purposes, Planning Interactive. This planning tool helps you write learning intentions and success criteria to match an overarching purpose. We recommend trying the Practice Interactive first.

Reference and Planning Resource: Go to Resources.Corwin .com/CreightonMathFormativeAssessment to access this resource.

Overarching Purposes and Guidelines for Writing LI and SC. This printable file provides a guide for considering purposes for lessons when writing learning intentions (LI) and success criteria (SC).

Recommendation 3: Keep your learning intention manageable. Narrow your focus to something that is attainable within a lesson or just a few lessons (understanding that a lesson may span more than one class period).

When Mr. Royce was planning the unit on proportions, he checked his state standards to be sure he would address the topic at the right level for his students. He saw that one of the standards said "Students use proportions to make comparisons between equivalent ratios." This broad goal defines a desired outcome for students by the end of Grade 7. In order to reach this goal, there are a number of mathematical ideas that are a part of meeting this standard that students need to understand, including what a ratio is, how to determine if ratios are equivalent in a problem, what a proportion is and why you might need one, and when it is appropriate or not appropriate to use a proportion. To write the learning intention for each lesson, then, Mr. Royce needed to focus in on an aspect of mathematical learning that is manageable to tackle in a lesson. (Note that this may mean more than one class period as oftentimes, a lesson will span multiple days.) The learning intention should communicate to the student what learning is most important and package it in a size that feels attainable to both teacher and students.

Recommendations 1 through 3 all focus on the part of your planning in which you articulate the important learning for your students. You may want to look at the guidelines resource available on the website to see how these recommendations appear in the guidelines.

Planning Resource: Go to Resources.Corwin.com/ CreightonMathFormativeAssessment to access this resource.

Evaluating and Refining LI and SC. This printable file provides guidelines for writing effective learning intentions (LI) and success criteria (SC).

A Note on Wording

In our experience, there are more and less effective ways to articulate a learning intention and success criteria that help you follow through on eliciting and interpreting evidence and providing formative feedback;

however, the perfect need not be the enemy of the good—worrying too much about getting perfect wording can actually get in the way of improvement. So it's worth saying a few words here on how to write a learning intention and success criteria more effectively rather than less.

In our work with teachers, we found that when teachers tended to write a learning intention starting with "Students will be able to . . ." (or other similar starters), we also found that their planning tended to focus on the acquisition or mastery of particular skills. Certainly not always but often enough. We also found that some teachers needed an alternative to using the word *understand*, because some districts considered the word too vague and required them to avoid its use in the learning goals for students. Instead, we encouraged teachers to frame their learning intentions in alternative ways:

- Start with "Today, we will learn _____" (and to avoid an overreliance on saying *how to* right after that). For example, "Today I will learn how surface area is related to area."
- Frame the central focus of the lesson as a question. For example, "Why do common denominators help when we're adding or subtraction fractions?" Proponents of the use of essential questions to frame learning may hear echoes of that approach here. However, essential questions are intended to be much broader in scope than is appropriate for a learning intention.
- Simply state the intended learning. For example,

  o "Ratios can compare parts to parts or parts to the whole."
  o "For three segments to form a triangle, the lengths of any two of the segments must add to a greater value than the length of the third segment."
  o "A graph is a representation you can use to show two pieces of information about something at the same time."

 Learning Resource: Go to Resources.Corwin.com/ CreightonMathFormativeAssessment to access this resource.

 Examples of Learning Intentions and Success Criteria. This database provides additional examples of learning intentions and associated success criteria.

These alternatives seemed to help the teachers we worked with move toward articulating the important learning, and different formats resonated with different teachers. Though each of them has merit, and you will see each of them used throughout this book, we would like to make a special case for the third format: Simply state the intended learning. Creating a learning intention statement requires a teacher to really think about what exactly is the intended learning, in a way that the other two formats don't necessarily require. The statement "Today, we will learn . . ." can be completed without necessarily saying precisely what that learning is; for instance, in our example earlier, exactly what is it about how surface area is related to area that the teacher wants students to understand? And asking a question can allow a teacher to inadvertently avoid having to explicitly articulate for himself or herself what the important learning is, unless the teacher answers the question for himself or herself ahead of time.

Some of the teachers initially balked at the idea of writing learning intention statements because they felt it would give away the point of a lesson, particularly if the lesson was structured as an inquiry lesson, in which students would go through a series of activities to discover a mathematical result. However, in our experience as teachers ourselves, we have found that students do not necessarily learn something simply by being told. Starting a lesson by saying, "Our learning intention for today is *for three segments to form a triangle, the lengths of any two of the segments must add to a greater value than the length of the third segment*," will not cause students to lean back in their chairs and take the rest of the class off because they now understand the Triangle Inequality. Instead, we contend that it makes more clear to students what it is they are going to try to learn, and that over the course of the lesson, as you clarify and revisit this learning intention, it actually serves to solidify students' learning. It may even pique their curiosity: "Is that really true? Why?"

There certainly may be some lessons in which stating the intended learning up front actually does a disservice to the lesson. Fortunately, there are a variety of ways to introduce a learning intention to students, and not all of them occur at the very start of a lesson, nor do all of them involve presenting the learning intention in its entirety all at once. Sometimes it may make sense to have students explore a mathematical idea first before homing in on what the learning intention will be. In other cases, it may make sense to provide a more bare-bones version of a learning intention, then gradually add to it, and flesh it out in more detail as a lesson progresses. A learning intention need not be viewed as a static statement, but rather it can be seen as something that evolves with the students as their understanding evolves. We take up a deeper discussion of these uses of learning intention in Chapter 5, "Developing Student Ownership and Involvement in Your Students." Meanwhile, suffice it to say that it does matter, to a certain degree, how you articulate your learning intention, but that there is not necessarily one right way to do so, so we encourage you to try out a format, perhaps more than one, to find what feels comfortable to you. While we are fans of the statement format, we have seen other instances of teachers using the "Today we will learn . . . " format and the question format quite effectively. And you may find yet another format not described here that works best for you.

Demonstrating Reaching the Learning

As you consider the success criteria that will indicate the extent to which students are reaching the learning intention, there are also three recommendations to consider, collected here:

> Recommendation 4: State the success criteria in terms of things that students can say, do, or produce, so that you and they can observe the evidence that learning is taking place.

> Recommendation 5: Decide what would constitute sufficient evidence—to you and to students—to indicate that students' learning was progressing toward the learning intention. Two or three success criteria are often enough.
>
> Recommendation 6: Align your success criteria to the learning intention so that the success criteria do in fact provide evidence of your stated learning intention and not some other learning.

After the incident with Evan and Alison, Mr. Royce talks with Ms. Dollard down the hallway. He learns that she has been listing for students a set of criteria that describes what students will be able to do or say or produce in their work if their learning is successful. Ms. Dollard is concerned about those students she has, like Evan, who can get the right answers much of the time but don't necessarily fully understand the material, so she makes sure she presents several success criteria, at least one of which asks students to demonstrate a skill or procedure, and—because she doesn't trust getting the right answer as convincing evidence that a student understands the mathematical ideas—at least one of which asks students to describe something about the underlying concept.

> Recommendation 4: State the success criteria in terms of things that students can say, do, or produce, so that you and they can observe the evidence that learning is taking place.

Because the success criteria are the indicators of reaching the learning intention, they must be framed to point toward different ways that students can demonstrate their understanding. The "Evans" who exist in any classroom may be adept at getting right answers but have no real substantive understanding of what they're doing (though the teacher's choice of mathematics tasks for students may sometimes inadvertently allow this to occur), so correct answers alone are rarely a sufficient indication that a student has fully mastered a concept.

Ms. Dollard showed Mr. Royce this example:

LEARNING INTENTION AND SUCCESS CRITERIA

Learning Intention: Understand how a graph can show how a quantity changes over time.

Success Criteria:

1. I can use the information in a line graph to make predictions.

2. I can use pictures, symbols, or words to show what the graph tells me about the data.

The highest-priority learning for this lesson is developing an understanding of how a graph shows change over time. If students are successfully making progress toward this learning, the teacher—and the students—will look for two particular things during the lesson: the extent to which students can use the information in the graph to make predictions and the extent to which students can describe the information about the data that the graph provides. The teacher will then plan the lesson (or modify an existing lesson plan) to ensure that he or she can gather evidence of each of these things.

It is important to note that determining *the extent to which* students have reached the learning intention is not necessarily the same as gauging *whether* students have reached the learning intention. From a summative assessment perspective, a teacher seeks to learn how well a student has mastered the material that has been presented; from a formative assessment perspective, a teacher seeks to learn how the student is making sense of the material so the teacher can make adjustments to instruction to better reach the student.

> **Recommendation 5: Decide what would constitute enough evidence—to you and to students—to indicate that students had reached the learning intention. Two or three success criteria are often enough.**

For Ms. Dollard's learning intention, two success criteria were needed. Meeting only one of the success criteria would certainly show progress toward reaching the learning intention but would fall short of fully reaching it. For example, if a student could use the information in a graph to make predictions but not describe what the graph tells about the data, the student might be able to apply a procedure of reading a graph for a single point but not see the big picture of change over time (which is specified in the learning intention). Similarly, if a student could describe what the graph tells about the data but not be able to use the graph to make a prediction, the student might see the big picture of long-term change but be unable to interpret specific instances, such as comparing how something changed from t seconds to v seconds.

On the other hand, Mr. Royce wondered if two success criteria were enough. He asked Ms. Dollard whether she's worried she'll overlook something important. She replied that her purpose in listing the success criteria is not to measure every mathematical skill or understanding that might arise in the lesson but to focus on the criteria that she feels are key to showing her that her students have reached or are on their way to reaching her learning intention. "And," she added, "if I happen to overlook something on one day that I perhaps should have included in my success criteria, I can always add it in the next day or on a subsequent lesson and look for it then. I'm trying to teach my students to pay attention to these as well, so I want to keep the number of success criteria manageable for both them and me to pay attention to."

> Recommendation 6: Align your success criteria to the learning intention so that the success criteria do in fact provide evidence of your stated learning intention and not some other learning.

After talking with Mrs. Dollard, Mr. Royce came up a new learning intention:

LEARNING INTENTION

What We'll Learn: Why you can set up proportions in different ways.

He then thought about what he wanted his students to be able to do or say during the lesson. He knew he wanted them to actually set up a proportion in several different equivalent ways, to solve each proportion for a missing value, to have more than one way to solve a proportion, and to justify why different proportions were equivalent.

Earlier in the chapter, we mentioned that while success criteria are tangible or observable, not all things that are tangible or observable in the classroom are necessarily success criteria. Mr. Royce has identified several potential criteria that he hopes his students will do as part of the lesson, but not all of them align with the purpose of the learning intention. For example, both solving proportions for a missing value and having more than one way to do so are important mathematical skills, but they do not provide evidence that students understand why they can set up proportions in different equivalent ways. Success with the remaining potential criteria would suggest some success with the learning intention, so they should serve as Mr. Royce's success criteria.

Note that the list of success criteria is not synonymous with the list of lesson activities or lesson agenda. As in Mr. Royce's case, students may be doing more things during the lesson than are listed in the success criteria; he still will ask them to solve their proportions, for example. The purpose of the success criteria is to pinpoint for the teacher and the students which of the things they're doing that will serve as indicators that students are reaching the learning intention.

In the end, he decided on two success criteria:

SUCCESS CRITERIA

What We'll Look For:

- *You can set up proportions in more than one way.*
- *You can justify why those ways result in equivalent proportions.*

He thought about whether this was enough evidence (Recommendation 5), and he decided that this felt sufficient to him.

A Note on Common Pitfalls to Avoid with Success Criteria

As we worked with teachers, it was the articulation of success criteria that felt most comfortable initially to many of them. Success criteria focus on tangible things that students do and felt familiar to both write and to present to students. However, there were several common pitfalls that arose that we describe briefly here, using Mr. Royce's learning intention and success criteria:

LEARNING INTENTION AND SUCCESS CRITERIA

What We'll Learn: Why you can set up proportions in different ways.

What We'll Look For:
- You can set up proportions in more than one way.
- You can justify why the two different ways give you the same answer.

---

1. **Success criteria can be too specific.** This pitfall can occur particularly when the success criteria relate too specifically to the lesson's activities. For example, if Mr. Royce wrote, "You can find the missing values in the problems on the handout," or "You can explain how you solved the problem in the lesson introduction," he would be providing success criteria that were too specific to the completion of a particular task. He would have focused on completing parts of the lesson activities, rather than stepping back to look across the lesson activities and describe an indicator that the lesson activities collectively contribute to.

2. **Success criteria can end up sounding like the lesson agenda.** "I can complete the problems on page 25," "I can share my answers with a partner," and "I can participate in the whole group discussion" are not success criteria. They are indeed all observable, and students may indeed have to do these things during the lesson. However, they focus on the lesson activities, rather than providing indicators of the important learning. And success criteria generally do not need to be met in any particular order. Success criteria need to help students gauge the effectiveness with which they are reaching the learning intention and might happen at different times for different students.

3. **Success criteria can end up turning into a long list, if they include all the necessary prerequisites to the lesson's idea.** For example, for Mr. Royce's learning intention, a too-long list of success criteria might look like:

What We'll Learn: Why you can set up proportions in different ways.

- *I can identify a ratio in a problem.*
- *I can write the ratio in different ways.*
- *I can explain how I know that two ratios are equivalent.*
- *I can set up a proportion using two equivalent ratios.*
- *And so on.*

While each of these things is important to understand and be able to do as a prerequisite to reaching the learning intention, they are not the most directly related to the learning intention. Success criteria help maintain the focus of the learning intention.

4. **Success criteria are not always "leveled."** In trying to make sense of success criteria, some of the teachers in our professional development program initially wrote them so that all students could meet the first one, some students could meet the second one, and only a few students could meet the third "challenge" success criteria. The success criteria are not intended to provide differentiation in a lesson, nor do they need to be written to ensure that all students can meet at least one of them. Their primary purpose is to help convey to students what the indicators are of reaching the learning intention, so it may be useful to think of them as describing the meets expectations part of the rubric.

Some readers may be feeling at this point like they are caught in the middle of the story "Goldilocks and the Three Bears," hoping to figure out how to write success criteria that are "not too this" or "not too that" but "just right." From our experience, these pitfalls are common ones and just part of learning how to generate and use success criteria. We encourage you not to be daunted by these pitfalls but rather just to be aware of them, and do your best to recognize them if they occur.

TEACHERS' VOICES

Thinking about creating learning intentions and success criteria was a bit stressful for me at first since I just didn't feel comfortable with how effective I was being in writing them. It was eye-opening for me to realize that if I couldn't put into words what I wanted students to know and understand, then how could I have a successful lesson that would move learning forward effectively? This is where it was better to use different information than just what the textbook described and look closer at the goals and objectives I have been working hard to curriculum map through the Common Core State Standards. When I focused

(Continued)

(Continued)

on what students need to know and learn and analyzed the text more carefully, I was able to look at the text as more of a resource that would guide instruction than being the entire instruction itself. This really built my confidence with writing learning intentions and success criteria, and now I can't imagine ever teaching a lesson without telling students what it is I expect them to learn and how I am going to go about doing that. Don't we all love to know that ourselves as adults? There is just no need to leave students in the dark when it comes to their own education.

Consider Your Students

As we've noted throughout this chapter, when students understand what they're trying to learn and what successful attainment of that learning could look like, they can become much more proactive in evaluating their own learning and letting you know what they need. This points to our final recommendation for writing a learning intention and success criteria.

> Recommendation 7: Write the learning intention and success criteria in language that is understandable to your students. If you need to introduce mathematical academic language, spend some time with your students clarifying what that language means.

Part of learning mathematics is learning the academic language of mathematics, so it is quite appropriate to include unfamiliar mathematical academic language in your learning intention and success criteria particularly when the lesson will involve developing a stronger understanding of that language. In Chapter 5, "Developing Student Ownership and Involvement in Your Students," you'll read more about some strategies for sharing a learning intention and success criteria with your students and ensuring that they understand what they are to be learning.

 Planning Resource: Go to Resources.Corwin.com/ CreightonMathFormativeAssessment to access this resource.

 Guidelines for Writing LI and SC. This printable file provides steps to help you write, evaluate, and refine learning intentions (LI) and success criteria (SC).

Classroom Resources: Go to Resources.Corwin.com/ CreightonMathFormativeAssessment to access these resources.

 Framework for Using LI and SC with Students. This printable file describes a way to help students understand and use learning intentions (LI) and success criteria (SC).

 Sharing Strategies. This printable file describes strategies for sharing learning intentions and success criteria near the beginning of a lesson, including the "Clarification Strategy," which provides a focus on clarifying any special academic language.

 Reference Resource: Go to Resources.Corwin.com/ CreightonMathFormativeAssessment to access this resource.

 Chapter 2 Summary Cards. These summary cards include the characteristics of learning intentions and success criteria, sentence starters for writing learning intentions and success criteria, and overarching lesson purposes.

Your Role During the Lesson: Sharing and Using Learning Intentions and Success Criteria

Look again at the Formative Assessment Cycle on page 54. Notice that the learning intention and success criteria are a focus in several places in the cycle and important in others.

Once the lesson begins, your focus shifts to sharing and making use of a learning intention and success criteria throughout the lesson. There are two parts to your role in implementing these formative assessment practices related to learning intentions and success criteria.

1. Your own use of the learning intention and success criteria during the lesson, including how you share them with students

2. Your efforts to develop your students' use of the learning intention and success criteria during the lesson

We touch on both of these parts here, but a more in-depth discussion of your use of the learning intention and success criteria during the lesson is included in Chapter 5, "Developing Student Ownership and Involvement in Your Students," after we've addressed other critical aspects in Chapters 3, "Gathering, Interpreting, and Acting on Evidence," and 4, "Providing and Using Formative Feedback."

Your Own Use of Learning Intentions and Success Criteria

The learning intention and the success criteria provide the backbone for other formative assessment practices when they are used in ways that support formative assessment. In brief, this means two things for you:

Figure 2.5 The Evolving Formative Assessment Cycle (with learning intention and success criteria)

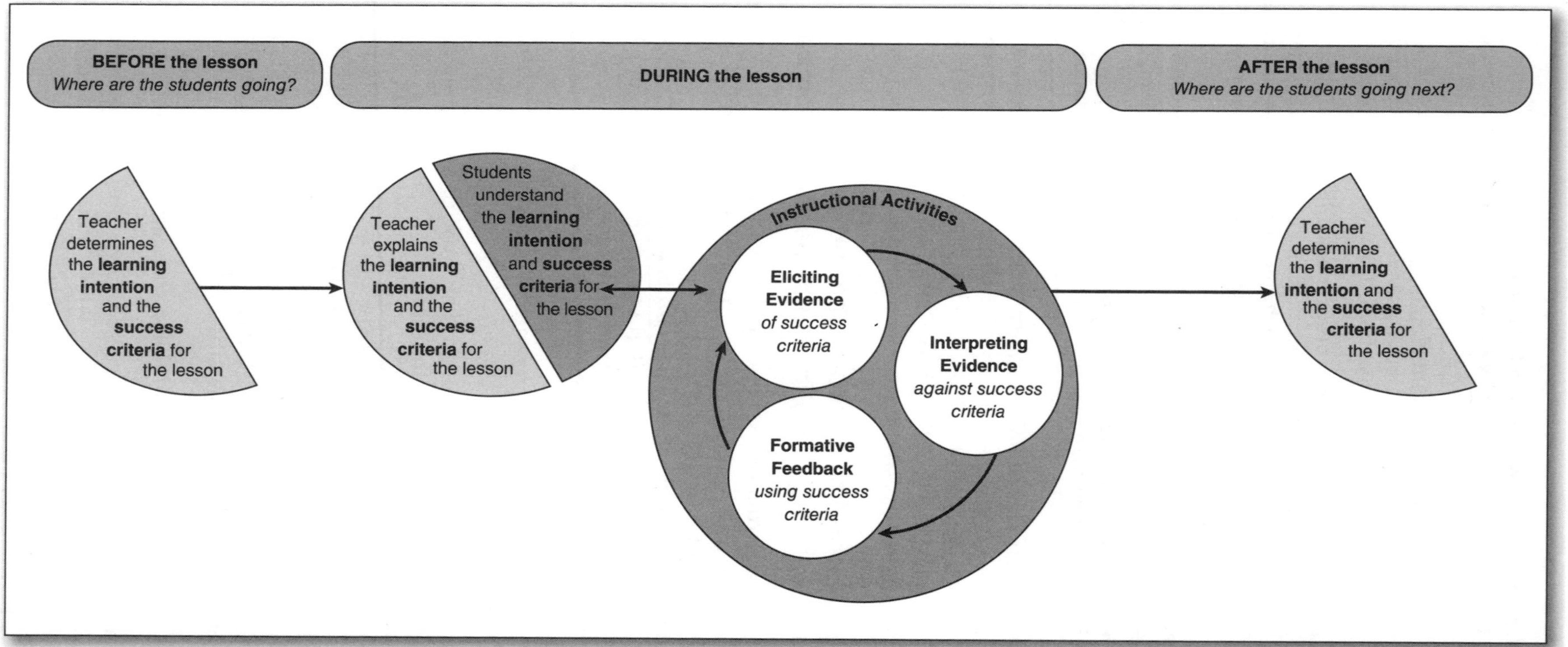

- Sharing the learning intention and success criteria with your students in ways that help them make sense of them
- Revisiting the success criteria during a lesson either to clarify them further for students, to pause to take stock of where students are in their learning, to help students consolidate their learning so far, or to provide feedback to students.

When you gather evidence of how students are doing, the success criteria provide the lens through which you interpret the results; when you provide formative feedback to the students to help them see what they are doing correctly and where they need to improve, referring to the success criteria provides a concrete way for students to see how your feedback is helpful to them. As you read more about these critical aspects in the next chapters, you will see how foundational the learning intention and success criteria are to all parts of the formative assessment cycle. For now, it's important to know that they provide a foundation for you during instruction to keep you focused on the important learning for your students and to serve as a basis for evaluating that learning while instruction is underway.

Several techniques have been developed that can help you share and revisit your learning intention and success criteria with your students. We come back to this idea in Chapter 5, "Developing Student Ownership and Involvement in Your Students."

 Classroom Resources: Go to Resources.Corwin.com/ CreightonMathFormativeAssessment to access these resources.

 Framework for Using LI and SC With Students. This printable file describes a way to help students understand and use learning intentions (LI) and success criteria (SC).

 Sharing Strategies. This printable file describes strategies for sharing LI and SC near the beginning of a lesson.

 Revisiting Strategies. This printable file describes strategies for revisiting LI and SC in the midst of a lesson.

 Wrapping-Up Strategies. This printable file describes strategies for revisiting LI and SC to help close or wrap up the lesson.

 Teacher Summary Cards for the Strategies. This printable file provides large-sized cards that provide step-by-step summaries of teacher moves for the strategies described earlier.

Reference Resource: Go to Resources.Corwin.com/ CreightonMathFormativeAssessment to access this resource.

Chapter 2 Summary Cards. This printable file includes the steps of the Framework.

Classroom Material: Go to Resources.Corwin.com/ CreightonMathFormativeAssessment to access this resource.

Student Summary Cards for the Strategies. This printable file provides large-sized cards with step-by-step summaries of student moves for the strategies described earlier. This can be distributed to each student or pairs of students for use when learning about a strategy.

Learning Resource: Go to Resources.Corwin.com/ CreightonMathFormativeAssessment to access this resource.

Images of Posted Learning Intentions and Success Criteria. This file provides images of the various ways teachers have posted the learning intention and success criteria for reference and use during a lesson. You will read more about posting learning intentions and success criteria in Chapter 7.

Supporting Your Students' Use of Learning Intentions and Success Criteria

While you clearly have a role in creating and using learning intentions and success criteria in your classroom, full implementation requires that your students interact with them as well. The sixth-grade teacher at the beginning of this chapter, Mrs. Daniels, talked about "starting the journey with the destination in mind." She found that when she shared the learning intention and success criteria with her students before the journey began, they were much more willing to follow her along the way to the destination, and she saw her students starting to engage in math class in a new and gradually more involved way. Starting the journey with the destination in mind involves helping students answer the first question of self-regulating learners: "What goals am I aiming for (in my learning)?" (See Figure 2.6.)

Gaining greater clarity about the goals they are aiming for in their learning allows students to gradually start to use the learning intention and success criteria in several important ways throughout a lesson:

- To maintain focus when there are many different mathematics ideas or skills at play during a lesson
- To self-assess their learning against the success criteria and the learning intention

- To make sense of feedback they have received
- To provide feedback to peers

Figure 2.6   Thinking Like a Self-Regulating Learner

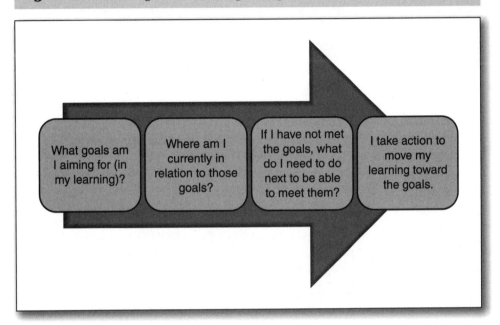

Many of the teachers we worked with initially were reluctant to devote much class time to sharing or revisiting success criteria or to helping students learn to do so, primarily because of the demands of their curriculum and the pressure to address all the necessary topics at their grade level. However, over time, they started to see a value in it that was not apparent at first. Part of that value rested with their struggling students. Students who are not performing well may often have difficulty pinpointing the important learning in a lesson and think they are required to do more than they actually are, because they struggle to prioritize the many different mathematics ideas or skills that arise during a lesson. As a result, they may end up doing more work, or just different work, than is desired (Wiliam, 2011). With all students, but especially for those students struggling with a math concept, time spent clarifying what good quality work looks like is time well spent because it can help them more fully understand what meeting any one of the success criteria could look like. For example, Mrs. Daniels might pull several examples of student work—in the moment, if the work is good, or anonymous work from another class—and have her students discuss the differences among the samples. Or using her own success criteria, she occasionally might pause the class to look at an example of student work that meets one or more success criteria and discuss with her students why. She would choose the examples to highlight specific points she wanted them to note. As students gain clarity about what it means to reach a learning intention, they are better able to provide quality responses in their own work. Ideally, as students become more self-regulating, they will turn to the success criteria themselves as they consider whether they are successfully learning what you intend. Students might see the destination at the beginning of the journey, but without periodic checking, they may not realize that they've begun to veer off the road. As they learn to self-assess using the success criteria, they can use the results along with the learning intention to guide themselves back toward the destination—putting their energies where they can do the most good.

To support students in their use of learning intentions and success criteria, we have developed a lesson to help them understand why learning intentions and success criteria are important. You will see more about this in Chapter 5, "Developing Student Ownership and Involvement in Your Students."

 Classroom Resource: Go to Resources.Corwin.com/ CreightonMathFormativeAssessment to access this resource.

 Introducing Students to Learning Intentions and Success Criteria. Resources for supporting your students in their use of learning intentions and success criteria are included in Chapter 5; however, you may want to take a look now at an introductory lesson to help students understand the purpose of learning intentions and success criteria. Look for Chapter 5 resources.

■ CONCLUSION

In this chapter, we have discussed what learning intentions and success criteria are and how they function within a formative assessment context. Together, a learning intention and success criteria provide the basis for students to understand what is the important learning at any given time and what successful learning looks like. While the learning intention helps students determine "What goals am I aiming for (in my learning)?" the accompanying success criteria spell out what it looks like when a student's learning is on track and helps the student determine "where am I currently in relation to those goals?" However, in order for a student to gauge where he or she is in relation to those goals, a student needs to compare his or her work to the success criteria. This work—in various forms—serves as evidence of the student's learning.

Recall that the learning intention and success criteria are foundational to implementing the other aspects of formative assessment. In this chapter, we offered recommendations for writing and learning intentions and success criteria that will help you support the other formative assessment practices. We also briefly talked about using learning intentions and success criteria with students, and we revisit a deeper discussion of this in Chapter 5, "Developing Student Ownership and Involvement in Your Students," when you are more familiar with "Gathering, Interpreting, and Acting on Evidence" (Chapter 3) and "Providing and Using Formative Feedback" (Chapter 4).

In the next chapter, "Gathering, Interpreting, and Acting on Evidence," we look at another critical aspect of formative assessment that helps both students and teachers address the second question in our self-regulation model: "Where am I currently in relation to those goals?"

Recommendations in This Chapter

1. Consider the mathematical understanding you want students to gain as a result of being able to perform certain procedures.

2. Focus on the highest-priority learning for students to attain—or to make progress toward—during the lesson.

3. Keep your learning intention manageable. Narrow your focus to something that is attainable within a lesson or just a few lessons (understanding that a lesson may span more than one class period).

4. State the success criteria in terms of things that students can say, do, or produce, so that you and they can observe the evidence that learning is taking place.

5. Decide what would constitute sufficient evidence—to you and to students—to indicate that students' learning was progressing toward the learning intention. Two or three success criteria are often enough.

6. Align your success criteria to the learning intention so that the success criteria do in fact provide evidence of your stated learning intention and not some other learning.

7. Write the learning intention and success criteria in language that is understandable to your students. If you need to introduce mathematical academic language, spend some time with your students clarifying what that language means.

Reference Resources: Go to Resources.Corwin.com/
CreightonMathFormativeAssessment to access these resources.

Chapter 2 Summary Cards. This printable file provdes Summary Cards that include recommendations from this chapter.

Formative Assessment Recommendations. This printable file provdes a summary of all the recommendations in this book in a printable (letter-sized) format.

RESOURCES ■

The following sections present some resources that can help you implement the recommendations. Each of these resources is referenced in earlier sections of the chapter, but here we provide a consolidated list. All resources can be found at Resources.Corwin.com/CreightonMath FormativeAssessment.

Learning Resources

These resources support your learning about formative assessment practices related to learning intentions and success criteria:

- **Lesson Purposes, Practice Interactive** provides practice considering purposes for lessons when writing learning intentions and success criteria; it also helps you learn how to use the "Lesson Purposes, Planning Interactive" listed in the planning resources.
- **Examples of Learning Intentions and Success Criteria.** Although writing learning intentions and success criteria is dependent on many factors including curricula, unit goals, and data from previous lessons, many teachers have found reviewing examples helpful. This searchable database includes the examples used throughout this book as well as additional examples.
- **Images of Posted Learning Intentions and Success Criteria** shows various ways teachers have posted the learning intention and success criteria for reference and use during a lesson. You will read more about posting learning intentions and success criteria in Chapter 7.

In addition, videos are available to support your learning about and use of learning intentions and success criteria.

Reference Resources

These resources summarize key ideas about formative assessment practices related to learning intentions and success criteria:

- **Chapter 2 Summary Cards** provide index-sized Summary Cards for the following:
  - **Key Characteristics** of Learning Intentions and Success Criteria summarizes the characteristics described in this chapter.
  - **Learning Intentions and Success Criteria Starter Statements** includes a list of sentence starters based on using *understand* as the basis of the learning intention.
  - **Purpose Categories for Writing LI and SC** lists the lesson purposes described more fully in "Overarching Purposes and Guidelines for Writing LI and SC."
  - **Evaluating and Refining LI and SC** reproduces the information in the "Evaluating and Refining LI and SC" planning resource.
  - **Framework for Using LI and SC With Students** summarizes ways to help students understand and use the learning intention and success criteria during a lesson. The Framework is described in detail in the classroom resource listed on page 61.
  - **Recommendations for Learning Intentions and Success Criteria** includes all the recommendations from this chapter.

- **Formative Assessment Recommendations** is a printable PDF that includes all the recommendations from this book.

Planning Resources

These planning tools are resources to support your lesson planning when writing learning intentions and success criteria for your mathematics lessons:

- **Lesson Purposes, Planning Interactive** provides a structure (similar to the "Lesson Purposes, Practice Interactive") for selecting purposes for lessons within a unit.
- **Overarching Purposes and Guidelines for Writing LI and SC** is a printable guide for considering purposes for lessons when writing learning intentions and success criteria. It can also be used as a reference when using the "Lesson Purposes, Planning Interactive."
- **Guidelines for Writing LI and SC** includes a step-by-step process for writing learning intentions and success criteria. The guidelines at the end help you evaluate and refine your learning intentions and success criteria.
- **Evaluating and Refining LI and SC** provides questions designed to help you evaluate and refine your learning intentions and success criteria. These questions are also included in the "Guidelines for Writing LI and SC" resource.

Classroom Resources

These resources illustrate various classroom routines that you can use during instruction. Each routine provides a structure that you can use or adapt to routinize your practice around the use of learning intentions and success criteria.

- **Introducing Students to Learning Intentions and Success Criteria** is an introductory lesson to help students understand the purpose of learning intentions and success criteria. More about this lesson is included in Chapter 5.
- **Framework for Using LI and SC With Students** describes a way to help students understand and use the LI and SC during a lesson. It describes actions for both teachers and students to take at the start of the lesson, at midway points during the lesson, and at the end of the lesson.
- **Sharing Strategies** includes several strategies for use when introducing students to an LI and SC for the first time, including the Clarification Strategy mentioned in this chapter. More about these strategies is included in Chapter 5.
- **Revisiting Strategies** includes several strategies for use when revisiting an LI andSC throughout the lesson. More about these strategies is included in Chapters 3, 4, and 5.
- **Wrapping-Up Strategies** includes several strategies for use referring to LI and SC to wrap up a lesson. More about these strategies is included in Chapters 3 and 5.

- **Teacher Summary Cards for the Strategies** is a set of large-sized cards that provides step-by-step summaries of *teacher* moves for the strategies described earlier. More about these strategies is included in Chapter 5.

Classroom Materials

This resource can be copied or recreated to be used in your classroom.

- **Student Summary Cards for the Strategies** is a set of large-sized cards that provides step-by-step summaries of *student* moves for the strategies described with the Classroom Resources. These cards can be distributed to each student or pair of students for use when learning about a strategy. More about these strategies is included in Chapter 3, 4, and 5.

3

Gathering, Interpreting, and Acting on Evidence

Mr. Antonelli was about to begin a unit on adding and subtracting fractions with his sixth-grade math class. He usually started the unit with estimating sums, so he gave his class a preassessment that included some pairs of fractions to compare. A number of his students chose the correct answer from the choices *greater than*, *less than*, or *equivalent to*. Marissa had the following answers on her paper:

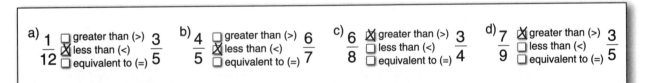

a) $\frac{1}{12}$ ☐ greater than (>) ☒ less than (<) ☐ equivalent to (=) $\frac{3}{5}$

b) $\frac{4}{5}$ ☐ greater than (>) ☒ less than (<) ☐ equivalent to (=) $\frac{6}{7}$

c) $\frac{6}{8}$ ☒ greater than (>) ☐ less than (<) ☐ equivalent to (=) $\frac{3}{4}$

d) $\frac{7}{9}$ ☒ greater than (>) ☐ less than (<) ☐ equivalent to (=) $\frac{3}{5}$

Mr. Antonelli looked at her answers and, since three of the four were correct, he determined that she probably had some good understanding of the relative sizes of fractions. He wondered about Problem c, though, since it seemed out of line with her other correct answers. So he decided to ask her to explain her thinking about why $\frac{6}{8}$ was greater than $\frac{3}{4}$. She replied, "$\frac{6}{8}$ is greater than $\frac{3}{4}$ because 6 times 8 is 48, and 3 times 4 is only 12, so that's how I thought that." "And is that how you compared $\frac{1}{12}$ and $\frac{3}{5}$ in

Problem a as well?" Mr. Antonelli asked. Marissa replied, "Yeah, 1 times 12 is 12, and 3 times 5 is 15, so $\frac{3}{5}$ is greater." As Mr. Antonelli looked at Problems b and d, he realized Marissa's faulty line of reasoning also resulted in the correct choice for both those problems as well.

Marissa's method for comparing fractions was consistently based on comparing the products of the numerator and denominator: She concluded that the larger product came from the larger fraction. The particular choice of fractions in each problem just happened to mask this misconception, except in Problem c. With more follow up, Mr. Antonelli discovered other students were using incorrect methods but happened to get mostly correct answers from them.

Teachers often look to correct answers as an indication of a student's understanding of a math topic. When the goal is to help students learn to solve certain kinds of problems correctly, using correct answers as information about or evidence of students' learning is entirely appropriate, though rarely sufficient. In Mr. Antonelli's case, he had collected some evidence of whether students had chosen the correct answer, but he had not collected any evidence of their approaches to answering the problem or their underlying conceptual understanding—how they were making sense of the mathematics and how that thinking led to their answers.

In this chapter, we discuss types of evidence, ways to gather it, and considerations for interpreting it to make subsequent instructional decisions. As we do so, we underscore the importance of collecting and interpreting a variety of kinds of evidence, with special attention to evidence of students' sense-making of the underlying concepts.

■ WHAT IS EVIDENCE?

Recall that, in the previous chapter, we discussed learning intentions that are focused on students' understanding of a concept and the success criteria that provide guidance about what to look for as tangible indicators that students are reaching the learning intention.

Let's look again at the "Thinking Like a Self-Regulating Learner" diagram, Figure 3.1 (p. 65), which summarizes the key questions and actions that self-regulating learners internalize and use to guide their learning.

When we refer to *evidence*, we mean anything that students say, do, or produce that allows you to *gauge the extent to which* students have met the success criteria. If students have fully met all the success criteria, they have likely also reached the learning intention; if students have only partially met the success criteria, there is probably an aspect of reaching the learning intention that they still need to work on. Success criteria and evidence are closely linked, and they can sometimes be confused as the same thing. Success criteria lay out the indicators for students of reaching the learning intention; evidence, on the other hand, consists of what students actually say or do, which teachers and students can then compare to the success criteria. As such, success criteria and evidence are key partners in formative assessment that help teachers and students determine what needs to happen next. In our evolving Formative Assessment Cycle, Figure 3.2, you can see how eliciting evidence and then interpreting the evidence against

Figure 3.1 Evidence in Self-Regulation

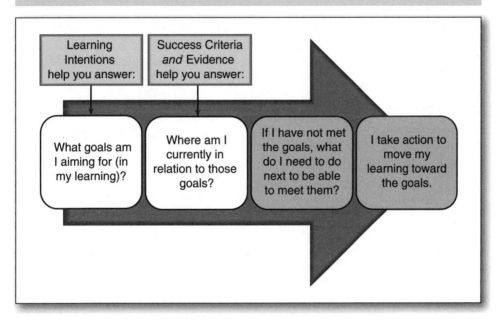

the success criteria fit into the part of the cycle that occurs during the lesson. Notice that we have included one question answered during the lesson by these evidence steps: Where are the students now? We talk later in the chapter about how teachers and students interpret the evidence and make decisions about appropriate next steps.

In the elicitation circle, you can see that students respond to the elicitation task. By *elicitation task* we mean any activity a student engages in that shows his or her thinking or understanding of mathematics. This elicitation task could have the sole purpose to elicit evidence, or it may be an instructional activity that also allows teachers and students to observe evidence of one or more success criteria.

Example: *Remember Ms. Jenkins's class from Chapters 1 and 2. Her success criteria for the lesson are*

1. *I can describe how a pattern of numbers continues.*

2. *I can describe how a corresponding visual pattern continues.*

3. *I can explain how the way a number pattern grows relates to the way the visual pattern grows.*

Ms. Jenkins's Warm-Up requires students to complete a table following a simple constant growth pattern. She gathers evidence of the first success criterion by looking at their responses and asking some students, chosen to give a good representation of the class, how they completed their tables. She finds that although most students are able to find the pattern, a few are unable to articulate how they

(Continued)

(Continued)

found their patterns. She makes a mental note to be sure they pair with a student with stronger articulation skills and to check on them during the first task.

During Exploration 1, students are looking for the patterns in the following table and set of figures and trying to connect the patterns:

1. Numbers:

X	1	2	3	4
Y	3	5	7	?

Figures:

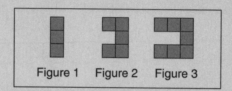

Figure 1   Figure 2   Figure 3

Ms. Jenkins gathers evidence for all three success criteria by observing students as they work in pairs and listening to their conversations. Occasionally she asks a pair to explain the patterns they find, and—after reminding them of the third success criterion—to explain how the patterns are connected. She interprets their responses to see the extent to which they are meeting the success criteria, listening for accurate descriptions of the patterns and for the students to explicitly connect the numbers in the visual growth to the numbers in the pattern growth (such as "2 more squares are added to make the next figure, and Y goes up by 2 each time" or "you start with the middle square and add the number of the figure above and below, so that's adding twice the figure number plus 1, and in the table you double X and add 1 to get Y, so you're doing the same thing").

As with many of the critical aspects of formative assessment, eliciting and interpreting evidence is something that is already an integral and common part of any teacher's work. All teachers gather and interpret information about, and make decisions about, students' learning and the necessary instruction. This occurs constantly through a myriad of interactions with students and numerous on-the-spot decisions that a teacher makes about where to focus, how to express an idea to students, where to go next, whether to spend more time with an idea—the list goes on and on. In this chapter, we focus on what it is about the eliciting and interpreting of evidence that is key to implementing formative assessment effectively. *Eliciting evidence for formative assessment* allows us to rethink the gathering of evidence in two specific ways: an awareness (and perhaps expansion) of what exactly we consider to be sources of evidence and being deliberate about linking the evidence to the learning intention and success criteria to maintain the focus on what you want your students to learn. *Interpreting evidence for formative assessment* points to one of a set of possible instructional moves, sometimes based on making in-the-moment

Figure 3.2 The Evolving Formative Assessment Cycle (with evidence)

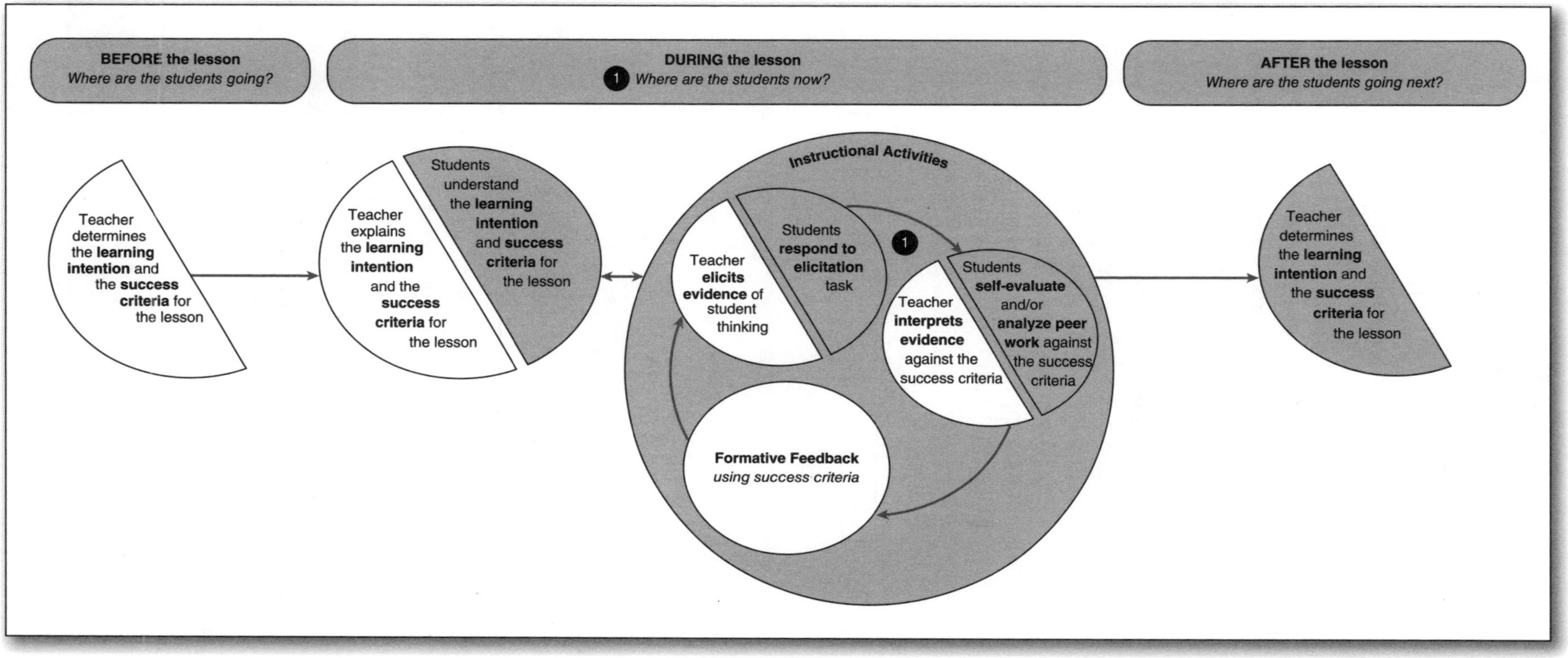

BEFORE the lesson
Where are the students going?

1 DURING the lesson
Where are the students now?

AFTER the lesson
Where are the students going next?

Teacher determines the **learning intention** and the **success criteria** for the lesson

Teacher explains the **learning intention** and the **success criteria** for the lesson

Students understand the **learning intention** and **success criteria** for the lesson

Instructional Activities

Teacher **elicits evidence** of student thinking

Students **respond to elicitation** task

Teacher **interprets evidence** against the success criteria

Students **self-evaluate** and/or analyze peer **work** against the success criteria

Formative Feedback *using success criteria*

Teacher determines the **learning intention** and the **success criteria** for the lesson decisions during the lesson and sometimes based on making decisions after the lesson is done.

Sources of Evidence

A commonly recognizable source of evidence is the tangible artifacts that students *produce*—homework, problems solved in class, or other written work they have done. These artifacts are central to seeing how a student is progressing in class. However, these examples are evidence of completed work, and in order to learn how students are making sense of mathematics ideas, it's important to also gather evidence of students' thinking while their work is in progress. So our definition of evidence also includes what students *say*—comments during class discussion, questions a student asks, or discussion you overhear when the student is working with a partner—and what students *do* when working on a problem. For example, observing ways that a student builds a pattern with color tiles can shed light on how they are mentally organizing the important quantities in the pattern. Consider this example, for which students are asked to identify a pattern and create the next figures in the sequence.

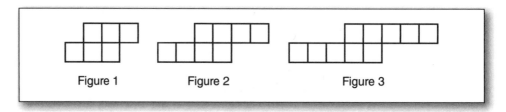

Two students might think about this pattern in different ways. Student 1 builds each figure by thinking of it as two rows of tiles that are growing in opposite directions.

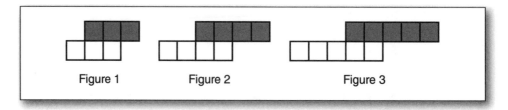

This student sees the pattern growing in terms of two identical quantities—the top row and the bottom row of squares. The student might describe this pattern by saying, "The total number of squares is double the top row. Each row grows out by one more square each time."

In contrast, Student 2 builds each figure by thinking of it as a central square that grows "arms."

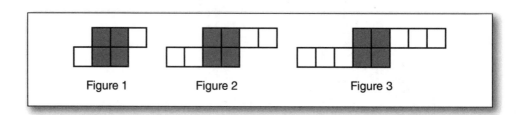

This student sees a constant core of four tiles, with an additional tile added to each arm for each new figure. Observing what students *do* as they build the figures—whether looking over their shoulder or having them present their thinking to the class—is another valuable source of evidence about how they are making sense of the mathematics in the problem. This evidence is gathered while the student's work is in progress, and that evidence can inform how you shape further instruction. Understanding how these students think about the figures helps you understand what expressions they create, and knowing that they're thinking differently prepares you for how you might bring up the differences—and similarities—as part of the instruction.

> ### Mathematics Background: Expressions for Figure Patterns
>
> *Suppose you want to move on to creating expressions for the number of tiles in the nth figure. Different ways of envisioning a pattern can lead to different—but equivalent—algebraic expressions for the pattern. When looking at patterns, it can be valuable to note what you see that stays constant in each figure (if anything), as well as what changes, and how it changes, in each figure.*
>
> - *Student 1 sees two rows of tiles growing in opposite directions. If the variable is used to refer to the figure number below each figure, then the length of each arm will always be two more than the figure number (or $n + 2$). Since there are two arms, you double $(n + 2)$ to get $2(n + 2)$ to find the total number of tiles in any figure.*
>
> - *Student 2 sees a central square of 4 tiles that grows arms. The length of each arm is equal to the figure number, and there are 2 arms. So the expression for the total number of tiles can be written as $4 + n + n$, or $2n + 4$. Note that this is equivalent to $2(n + 2)$.*

COMMON CORE CONNECTION

The Standards for Mathematical Practice describe ways of engaging with mathematics, and as such, evidence of these practices are best found in the work and the thinking that students do, rather than in their final solutions. Looking for evidence of how students are making sense of the mathematics while their work is in progress can also provide opportunities to look for evidence of the math practices in students' mathematical thinking.

Aligning Evidence to Learning Intentions and Success Criteria

As noted, evidence and success criteria are key partners in the formative assessment process that help teachers and students determine what needs to happen next. The evidence that you gather needs to be clearly connected enough to the success criteria that you and your students can use the evidence to make judgments about the extent to which students are reaching the success criteria; we refer to this as having evidence that is *aligned* with the success criteria and therefore also the learning intention. To illustrate what this means, let's consider an example in which the evidence is *not* aligned with the success criteria or the learning intention for a lesson.

Mrs. Keller wants her eighth-grade algebra class to understand slope as a constant rate of change. During class, she has her students practice finding rates of change in word problems, graphing a corresponding line, and then finding the slope of the line, and at the end of class, she collects their papers to look over what they did.

Mathematics Background:
Slope and Constant Rate of Change

Linear relationships are an important topic in middle school mathematics. This topic focuses on patterns of numbers that increase by adding a constant amount each time. For example, if I earn $10 an hour for each hour I work, the relationship between the number of hours I work and the amount of money I earn is linear; it increases by adding $10 for each additional hour.

Linear relationships appear on a graph as a straight line. When students are learning about the slope of a line on a graph, they learn that the slope remains the same, no matter what part of the line you're looking at. This is because as one variable increases by a certain amount, the other variable also increases (or decreases) by a fixed amount; for every hour I work, my pay increases by $10. This rate of change remains constant, so the steepness of the line also remains constant on any part of the line. The slope of a line is equal to the rate of change, and it can be used as a numerical measurement of the steepness.

In this example, Mrs. Keller wants to develop her students' understanding of how slope indicates a rate of change that is constant, and therefore the graph of the situation will be a line. However, the activities she lays out and the evidence that she is able to gather from those activities focus on procedural skills: graphing the line and determining the slope. There is a mismatch between the learning she is trying to develop in her students and the information she collects about their learning. Some of her students may very well be starting to see the relationship between constant rates of change and the slope of an equation, but she cannot yet see that from the evidence she has collected, because the task she lays out for them does not provide her with the kind of evidence she is looking for.

Now let's look at an example in which the evidence is well aligned to the learning intention. In this example, Mrs. Keller identifies her learning intention and success criteria as the following:

LEARNING INTENTION AND SUCCESS CRITERIA

Learning Intention: In linear functions, there is a relationship between the slope and the rate of change in the two variables.

Success Criteria:

1. I can describe what "rate of change" means.

2. I can find the rate of change in a situation.

3. I can explain how rate of change is related to the idea of slope.

In this example, Mrs. Keller now has identified three key indicators she would like to see if her students are reaching the learning intention: generating a definition of rate of change, actually finding the rate of change in a problem, and then explaining its relationship to slope. Given these success criteria, she can now make sure she includes activities in her lesson plan that provide opportunities for students to give evidence of each of these criteria. We illustrate how she does that in the next section, Using Evidence in Your Classroom.

USING EVIDENCE IN YOUR CLASSROOM: ■
THE TEACHER'S ROLE

There are four important components to the use of evidence in your classroom, and each can be characterized by a central question, as shown in Figure 3.3.

Figure 3.3   Components of Using Evidence

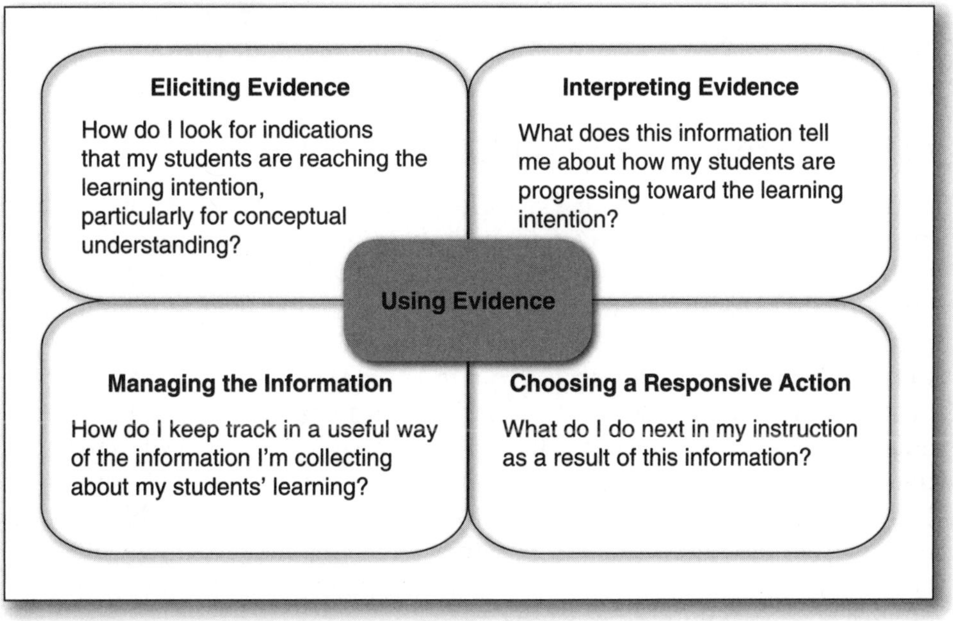

Eliciting Evidence

How do I look for indications that my students are reaching the learning intention, particularly for conceptual understanding?

Interpreting Evidence

What does this information tell me about how my students are progressing toward the learning intention?

Using Evidence

Managing the Information

How do I keep track in a useful way of the information I'm collecting about my students' learning?

Choosing a Responsive Action

What do I do next in my instruction as a result of this information?

In this section, we delve more into the use of evidence to develop answers to each of these questions. The students' role in eliciting and interpreting evidence, and how you can support them in that role, is discussed in the next section.

Eliciting Evidence of Meeting the Success Criteria

The first part of using evidence in your classroom, eliciting evidence, involves addressing the question, "How do I look for indications that my students are reaching the learning intention, particularly for conceptual understanding?"

Figure 3.4 Eliciting Evidence Component of Using Evidence

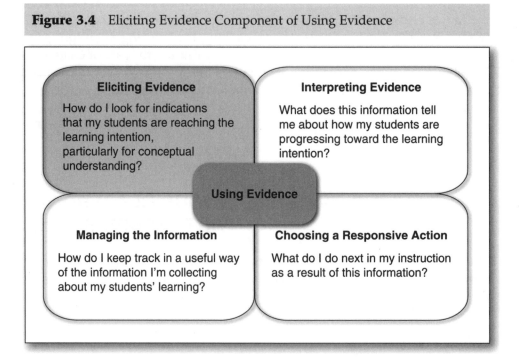

Sometimes, spontaneous opportunities arise in a math class to effectively elicit some valuable evidence of students' thinking, but eliciting evidence can be even more effective with some careful planning before the lesson.

Before the Lesson: Planning for Eliciting Evidence

Your lesson planning process should include consideration of the first question in Figure 3.4: "How do I go about looking for indications that my students are reaching the learning intention?" Different approaches can be taken to answer this question, but one approach that we found helpful is summarized in Figure 3.5. A teacher first identifies the key understanding that is the priority in the lesson. From that, he or she can then decide what he or she would consider as reasonable indicators that the students were gaining this understanding. The teacher can then figure out what evidence he or she needs to gather for those indicators and therefore what lesson activities need to take place to provide that evidence. Notice that the first two thought bubbles focus on identifying your learning intention and success criteria. The remaining bubbles provide the planning needed to elicit evidence within your lesson.

Figure 3.5 Creating Opportunities to Elicit Evidence

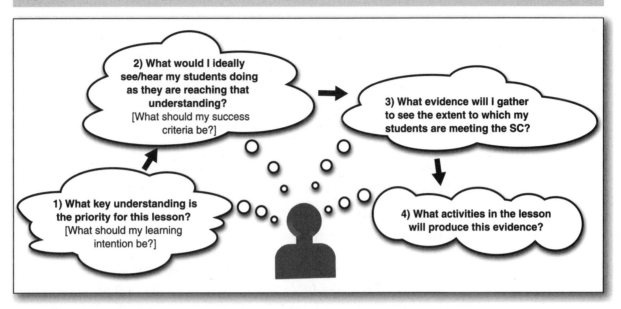

Using this approach, formative assessment practices become fluidly embedded within instruction—not a separate assessment activity that interrupts learning in order to occur. The clearer you can be about your purpose for gathering and interpreting the evidence—being much more specific than "I want to know how they did"—the more valuable the results of your evidence gathering will be, both to you and to the students. The following recommendations will help you in your planning for evidence to gain that clarity.

> **Recommendation 1: Create opportunities during the lesson to gather evidence related to each of the success criteria.**

This recommendation echoes the last two thought bubbles in Figure 3.5, and it is called out specifically as a recommendation because, while it is key to effectively using evidence, it can easily be overlooked.

One consideration with this recommendation is *what* evidence you will collect. This recommendation emphasizes that you create opportunities during the lesson to gather *evidence related to the success criteria*. This focus on the success criteria is analogous to the system used by many teachers for a writing process, in which several criteria are spelled out and a student's writing is assessed only in terms of those several criteria, not for every possible thing that could be corrected. It underscores the importance of having success criteria that point you where you want to go in the lesson, that are well aligned to the learning intention, and that are clearly articulated for yourself and for your students.

A Note on a Common Pitfall: Misaligning the Task with the Evidence

One common pitfall is the potential for misaligning the math task and the information you want to gather from it. This can commonly occur in two different ways. The first way this misalignment can occur is by posing a task that only provides evidence of students' ability to complete a procedure (*skill performance*) rather than providing evidence of their understanding (*sense-making or understanding of the key idea*). Mrs. Keller's first example is an example of this misalignment. In that example, her students practice finding rates of change in word problems, graphing a corresponding line, and then finding the slope of the line. Although she wanted a conceptual goal (connecting rate of change to slope), her tasks focused on procedural work (calculations and graphing) and only provided evidence of procedural skill.

Another important consideration is *how* you will collect the evidence. There are many strategies for gathering information about student thinking and skills that you can embed within lesson activities, for example,

- "I used to think . . . , but now I think . . . " exit ticket writing prompts.
- card sorts, in which information is written on cards, and students process the information in some way to sort the cards.
- "Now that you mention it . . . "—a discussion protocol for students in which they take turns talking with a partner about a particular topic or question for a certain amount of time as well as responding to what the partner said.
- response cards labeled A, B, C, and so on, that students can hold up for you to see, used with multiple-choice questions.
- rich questions that invite students to share their thinking.

Classroom Resources: Go to Resources.Corwin.com/ CreightonMathFormativeAssessment to access these resources.

 Elicitation Techniques. This printable file contains a list of additional strategies for eliciting evidence that can be found in the book *Mathematics Formative Assessment: 75 Practical Strategies for Linking Assessment, Instruction, and Learning* by Keeley and Tobey (2011). A brief description of each strategy is included.

 Elicitation Strategies. This printable file describes strategies to elicit evidence that you might find helpful.

 Framework for Using LI and SC With Students. This printable file describes a way to help students understand and use learning intentions (LI) and success criteria (SC).

 Revisiting Strategies. This printable file describes strategies for revisiting LI and SC in the midst of a lesson, including strategies for use when eliciting evidence related to the LI and SC during the lesson.

 Wrapping-Up Strategies. This printable file describes strategies for revisiting LI and SC to help close or wrap up the lesson, including strategies for use when eliciting evidence related to LI and SC during the lesson.

 Teacher Summary Cards for the Strategies. This printable file provides large-sized cards that provide step-by-step summaries of teacher moves for the strategies described earlier.

Classroom Material: Go to Resources.Corwin.com/ CreightonMathFormativeAssessment to access this resource.

Student Summary Cards for the Strategies. This printable file provides large-sized cards that provide a step-by-step summary of student moves for the strategies described earlier. This can be distributed to each student or pairs of students for use when learning about a strategy.

Different strategies will yield different kinds of information, so when choosing an activity that elicits evidence, pay careful attention that the strategy you choose provides the type of information you're looking for. For example,

- response cards can give you a quick sense of the class overall and can show you who is getting a correct answer, but they cannot show so easily how students are thinking about the problem.
- "I use to think ... but now I think ..." can give you information about an individual student's thinking and how it has changed during the lesson but cannot show so easily how well or how efficiently students can solve various kinds of problems.
- card sorts can provide you information on what connections students see between the various ideas on different cards but cannot show so easily where they are getting tripped up.
- if you listen in on several pairs of students as they use the "Now that you mention it ... " protocol, you can collect typical examples of things students are thinking, but you can't as easily get a sense of what is typical of the larger class.
- rich questions provide the opportunity for some students to explain how they think about the mathematics, but again, getting a sense of the whole class may be difficult.

Each strategy is well suited for providing some kinds of evidence but not others. Thus, the best approach is to incorporate more than one evidence-gathering strategy during a lesson to ensure you have more than one opportunity to collect the information you need.

A final consideration is *when* you will collect evidence during the lesson. There are many opportunities in a lesson when you could elicit evidence—too many, in fact, for you to be able to take advantage of every one. Identifying key places in the lesson allows you to be strategic about when you might elicit evidence of how your students are thinking about the mathematics. Leahy, Lyon, Thompson, and Wiliam (2005) describe a *hinge point* at which your lesson might take different paths, depending on your students' needs at that moment. This could be when they are working on individual practice: if students are performing, or responding to your questions about their work, as expected, then you would probably feel comfortable moving on to the next, new piece of learning. If you see difficulties arising, or hear students expressing one or more misconceptions or misgivings about the mathematics, you could choose a new experience with the same material. Similarly, you could ask a *hinge-point question* to a larger portion of the class (or all of it) when a task has been completed, to gauge which of the options you have prepared is most appropriate. Multiple-choice response cards or whiteboards allow all students to response simultaneously to a question that isolates particular misconceptions or barriers; then you can choose how to proceed, including dividing the class into smaller groups for different tasks based on their responses.

A Note on a Common Pitfall: Saving All Your Evidence Gathering for the End of Class

In response to the pressure you may feel to address all the content at your grade level, it may seem that you don't have time to build in opportunities to elicit evidence during the lesson; instead, you want to just get through the material and check for students' understanding at the end of the lesson. While this is a perfectly fine way to gather evidence for some lessons, it's likely that some rich opportunities are being missed during your lesson, especially if this is the only means you have to gather evidence. Time spent while the lesson is underway can save time later, because you can redirect students, and your own instruction, as you go—skipping parts they actually know while focusing on the parts they don't.

> **Recommendation 2: Consider from whom you need evidence. You might not need it from all students all the time on all parts of the lesson.**

Part of your planning to elicit evidence involves thinking about *from whom* you will elicit it. Not all opportunities to gather evidence need to include all the students in your class. Some evidence may be appropriate to gather from the whole class, such as using response cards to collect responses from all students on a hinge-point question between instructional tasks or collecting exit tickets from each student in which they summarize their understanding of a key idea from the lesson that day. Other evidence may be appropriate to gather from a sample of students in the class. For example, in deciding in the middle of a lesson whether to spend more time with a new representation you've introduced, you may have four or five students who serve as good indicators of the rest of the class for you—if they are ready to move on, the rest of the class is also likely ready to move on. Yet other evidence may be appropriate to gather only from one or two individuals, such as checking on the understanding of a student who you helped after school the previous day or of a student who got an introduction to today's lesson topic in his or her math support class prior to coming to your class.

Let's look at Mrs. Keller's planning process to illustrate this pair of planning recommendations.

Mrs. Keller has established her learning intention and success criteria:

LEARNING INTENTION AND SUCCESS CRITERIA

Learning Intention: In linear functions, there is a relationship between the slope and the rate of change in the two variables.

Success Criteria:

1. I can describe what "rate of change" means.

2. I can find the rate of change in a situation.

3. I can explain how rate of change is related to the idea of slope.

She now thinks about what evidence she would need to gather for this collection of success criteria, ensuring that none of them are overlooked. From a preassessment she gave the class, she knows that many of her students can accurately calculate the slope of a line. Some of her students have said that the phrase rate of change is familiar, but they are not clear how it relates to lines.

o *She focuses first on evidence for Success Criterion 1.*

Mrs. Keller thinks, "I want students to give me a description of what they think *rate of change* means both as we start the lesson and as we conclude the lesson, so I can see how their understanding of that phrase has changed, if it changes at all. I can have them do this either verbally or in writing. I can also listen for comments they make during class as they're working."

o *She focuses next on evidence for Success Criterion 3 because she sees an opportunity to combine eliciting evidence for Success Criteria 1 and 3.*

"As we get further into the lesson, I want to know how they think rate of change is related to slope. As they talk about their understanding of rate of change again either in discussion or in writing (or maybe both?), I'll need to build in some way to specifically ask them about this relationship."

o *Then she focuses on evidence for the remaining Success Criterion 2.*

"I'll need to see whether they can take a word problem and actually identify, in words, what the two variables are in the relationship in the problem and what the rate of change is between them. This can be in their written work."

Now that she has articulated the evidence she wants to be able to elicit, she can easily identify the appropriate lesson activities to create the opportunities to elicit that evidence.

o *She thinks first about the activity to provide evidence for Success Criterion 1.*

"I want to find out what they're thinking rate of change means before I start providing a definition, and I'll eventually want to build a definition from what they currently understand. So at the start of the lesson, I'll use several word problems to prompt a discussion about what rate of change means. I'll have students talk briefly in small groups first, then share their ideas with the whole class. The word problems will provide some specific examples of real-life situations in which students can talk about what the two variables are in each problem and how each variable is changing in relation to the other. They also need to be able to compare two word problems, so they can decide if one changes faster—that will bring out the rate of change idea in a concrete way. As they share their ideas, I can gather evidence of how they currently make sense of the idea of *rate of change*."

o *She decides to keep a record of students' ideas to revisit and refine later in the lesson.*

"Maybe I should collect their ideas on the board and keep them for reference. We can revise them together later in class. This will give me information about how their thinking is evolving (and *if* it is!) and will help them solidify their understanding."

o *She feels that, after this initial discussion, students need some time to work on some problems in small groups, so she turns her focus to activities to provide evidence of Success Criterion 2.*

"I want my students to identify the rates of change in various word problems, but they're going to need some help being able to do that. I'll give them a sentence frame to use as a structure for talking about the rate of change: 'For every _____, _____ increases/decreases by _____.' We'll do the first example together, so they can see how to use this. Maybe something like 'For every hour I work, my pay increases by $10.'"

"I'm also going to ask them to graph the relationship described in the word problem and calculate the slope of the line. I want them to simply start noticing that there's a connection between the rate of change and the slope of the line. They may not be able to tell me what it is yet, but I'd like them to start noticing there's a connection."

"Then as I walk around, I'll be looking for whether they're able to describe the rate of change and how they describe it. In fact, I should probably especially listen to Mark, Luis, Gemma, and Katy. Gemma and Mark are pretty good examples of my kids who need more digesting time for ideas. Katy needs concrete examples but can then put the ideas together, and Luis can usually understand conceptual ideas pretty quickly. They should give me a pretty good sense of the rest of the class."

o *She then decides to bring the class back together for a summary discussion about "How is your thinking about what 'rate of change' means the same or different from what you thought earlier?" to gather additional evidence about Success Criterion 1.*

"I'll have students come back together; and as a whole class, we'll discuss revisions to the list; and I'll write up any changes or new ideas that students have. I can also use this as an opportunity to make sure we're being precise in our language."

o *She plans to add the question, "What does the idea of 'rate of change' have to do with the slope of the line you graphed?" to see what evidence she can elicit for Success Criterion 3. She realizes the class may or may not be ready for this discussion, and she will need to decide in the moment during the lesson whether to go in this direction or not.*

"I may or may not get to the slope discussion question. If I hear them talking about a given change in one quantity corresponding to another given change in the other quantity, such as '1 hour more work means $10 more,' then they are probably ready to connect rise and run in the slope calculation to these changes and we can move on to that discussion. If I don't hear that, I may need to work with them to see rate of change that way. I'll find out as we get into the lesson."

o *She plans to conclude the lesson by having students complete a short individual writing prompt, called an "exit ticket," to gather evidence from each student individually about his or her understanding of the learning intention so far. She does not expect everyone to necessarily have reached the learning intention quite yet, but she wants to see where their thinking currently sits, to inform her planning for the next lesson.*

"I'm going to keep this short—just one question—because it needs to be quick for students to answer and not cumbersome for me to read through and analyze after the lesson! I'll have them complete a final exit ticket before they can leave with a prompt either saying, 'At this point, how do you think *rate of change* is related to the idea of *slope*?' if we've gotten that far in the lesson, or if not, I'll make the prompt be 'At this point, what do you think *rate of change* means?'"

In this example, Mrs. Keller has several examples of evidence: students' ideas about rate of change gathered at the start of class, their comments and written work in small groups, their contributions to the whole-class discussion, and their written response to the exit ticket. This evidence is more closely aligned to the learning target because the tasks and lesson activities chosen provide information about how students' conception of rate of change is developing further over the course of the lesson and how students are relating the ideas of rate of change and slope to each other.

It's noteworthy that in her lesson plan, the gathering of evidence is not separate from the lesson's activities; they are one and the same. Her choice of lesson activities—the initial discussion about rate of change, the small group work on the problems, and the ensuing whole-class discussion—are specifically chosen to maximize opportunities for students' learning by involving them centrally in describing their understanding of the relationship between slope and rate of change, rather than Mrs. Keller simply telling them what the relationship is. She may choose in a later lesson to give a brief lecture to provide more information, but because in this lesson she wanted to gather evidence of students' understanding of the ideas, she chose a different lesson structure and activities.

It's also noteworthy that when planning opportunities to gather evidence for the success criteria, any one type of evidence Mrs. Keller gathers does not need to address all the success criteria simultaneously. In each case, the evidence being gathered targets one or maybe two of the success criteria, so that by the end of class, *the collection of evidence provides information across all the success criteria* and can better inform her how well her students have met the learning intention and where their learning needs to go next.

Finally, notice that much of the evidence Mrs. Keller plans to collect is easy to gather and to interpret quickly. She doesn't need to record any of this evidence, but it will provide her with the information she needs to make in-the-moment decisions about her students' needs and what the next steps for the class will be.

A Note on a Common Pitfall: Interpreting Evidence Always for Individual Students

Recommendation 2 suggests choosing from how many, and in some cases from which, students you will elicit evidence. When you have evidence from all students, you may be tempted to look carefully at each student's response and to carefully assess each student individually. After all, you're trying to understand how your students are making sense of the mathematics content. To help keep things manageable for yourself, though, first be clear about your purpose in eliciting the evidence—what are you trying to find out about your students? For example, you could be wondering, "Which of my students had the most difficulty reaching the success criteria?" or instead "How much of the class is still unclear on the concept of ____?" You could also be wondering, "What kinds of difficulties are students still having with ____?" or "What do my students collectively understand so far about the learning intention?" If you intend to get a feel for the whole class, resist the urge to break it down to the individual student. We say more about how to interpret evidence for a whole class collectively in Recommendation 6.

Planning Resources: Go to Resources.Corwin.com/ CreightonMathFormativeAssessment to access these resources.

 Guiding Questions for Formative Assessment. This printable file provides a set of guiding questions to help you plan for eliciting evidence for different purposes as well as examples of techniques suited for a particular purpose.

 Formative Assessment Planner Templates. This printable file provides two versions of templates to help you create a formative assessment plan for a lesson. These templates will be revisited in later chapters so for now pay particular attention to the portions related to setting learning intentions and success criteria and planning for how and when to collect evidence.

 Reference Resource: Go to Resources.Corwin.com/ CreightonMathFormativeAssessment to access this resource.

 Chapter 3 Summary Cards. This printable file provides Summary Cards that include planning for eliciting evidence.

During the Lesson: Eliciting Evidence

Planning for eliciting evidence focuses on creating opportunities in a lesson to gather evidence of each of the success criteria and being thoughtful and intentional about from whom you decide to gather evidence. As you implement your lesson plan, you will undoubtedly face more decisions about eliciting evidence. The following recommendations can help you with those moments in which you might want to elicit more evidence from your students.

> **Recommendation 3: Elicit evidence about correct thinking and correct responses as often as you do about incorrect thinking and responses.**

Earlier in the chapter, we talked about how eliciting evidence is about answering the central question, "How do I go about effectively looking for indications that my students are reaching the learning intention, particularly for conceptual understanding?" This means looking for indications of what skills students are using correctly as well as where they are thinking in productive ways and developing accurate understanding of conceptual ideas. Yet too often in mathematics classrooms, we can get disproportionately caught up in what students do not understand and where they are going wrong. This can lead to an overemphasis on students' deficits in how we think about our students, how we structure their learning, and how we talk to them. This recommendation is about shifting that emphasis so that we give equal attention and airtime to students' successes and their accurate thinking.

Part of this shift involves looking first for examples of correct work or correct thinking, rather than first looking for deficits. In many instances, student mistakes have the seeds of correct thinking buried within them; they may be incorrectly applying thinking that was correct for a prior topic. For example, when multiplying fractions, some students start by (unnecessarily) finding a common denominator; they may be applying what they know (or partially know) about fraction addition and subtraction to the wrong operation. When students incorrectly multiply the area of a pyramid's base by its height to find volume, they might be thinking that the formula for a prism with the same base would provide the volume for the pyramid as well. When students incorrectly try to multiply f times a number in the function notation $f(3)$, they are likely applying what they know about notation for multiplication to a new use of that notation. In each example, there is something that the student does know that they are trying to apply to a problem that they may not know how to solve. The more you can look for those seeds of correct thinking, and build from them, the more opportunities you create to build your students' confidence that they *can* do mathematics and that they do have some successes, rather than broadly declaring, "I can't do math."

The other part of this shift is taking action to explicitly gather evidence of thinking that produces correct responses as well as thinking that produces incorrect responses. We've already seen, with Mr. Antonelli in the opening example for this chapter, how correct answers are not always an indication of correct thinking. Gathering additional evidence to verify that correct work is correct for the right reasons can go a long way toward enacting this recommendation. Teacher follow-up questions to a student response are another easy way to elicit more evidence about correct thinking. You not only benefit from gathering more information about a student's thinking, but the act of having to say more about his or her thinking helps the student solidify his or her thoughts and can help other students who may need more elaboration in order to make sense of the idea under discussion.

Because teacher follow-up questions are such a rich way to elicit evidence of thinking, both correct and incorrect, and because they are such an integral part of any classroom, we focus on questioning in this next recommendation.

> **Recommendation 4: Use questions (at least some of the time!) to learn more about students' mathematical thinking.**

As you saw in Recommendation 1, there are numerous ways to collect evidence during a lesson. Eliciting evidence during the lesson is partly a matter of enacting the plan you created before the lesson. However—as all teachers know well—once your lesson begins, your work with students can take a variety of sometimes unexpected twists and turns from what you envisioned while planning. While you can plan what *you* will say and do, you cannot necessarily plan how *your students* will make sense of what you say and do. You need some techniques that you easily can use, and adapt, depending on how students respond to your lesson activities. *Questioning* is one particularly versatile and helpful technique, with such varied and complicated use in the classroom that studies have been conducted to describe and clarify its use in the classroom.

Teachers use questions for a wide range of instructional purposes. For example, questions may be asked to get students oriented at the start of a problem, to assess their understanding, to refocus them when needed, to advance their thinking, or for classroom management purposes, to name a few (Driscoll, 1999; Student Achievement Division, Ontario Ministry of Education, 2011). When we focus on questions that best serve the purpose of gathering evidence of students' thinking for formative assessment, several characteristics of these questions become evident:

- They elicit a response that provides insight into how your students are thinking or reasoning about the mathematics, and they can't be answered with a single word, number, or short phrase.
- They provide evidence specifically *about the success criteria*, allowing you to gather significant information about the students' thinking relative to the tasks they are working on.
- They can be planned for ahead of time, or they can be spontaneous during instruction.
- They occur at several key points throughout the lesson: (1) as you are establishing the success criteria, (2) while students are working on a task, and (3) after students finish a task.

To fully implement formative assessment, students need to become comfortable with and accustomed to explaining their thinking in various ways, and questions are a great way to get at that thinking. Your purposes in asking questions will depend on where students are in the task they have been given.

- Prior to working on a task, you might ask students questions to determine what they think they are being asked to do in the task you've put in front of them and what mathematical ideas and prior learning they associate with that task. You may want to use the responses to those questions to identify which students may need extra help as they work. You may also provide a short review of content or clarify part of the task.
- While students are working on a task, you might ask them questions to learn more about their reasoning and their current understanding or to clarify their articulation of their ideas. Some students may appear to be completely off track, but when they explain their thinking, you may find that they are approaching the problem in a valid way, just differently from what you might have expected. If you discover a student is able to continue working, you might decide in the moment not to pursue more evidence or not to respond to the student; just step back and let him or her work.
- When students are finished with a task, you might ask them questions to clarify what strategies they used and identify the extent to which they understand their strategies, hear them reason

mathematically about the task, and determine how well their thinking connects with the content in the success criteria. Differences in students' strategies and solution methods can highlight how students are thinking about the content as well as provide an opportunity to discuss different approaches to a problem. Students who were struggling can benefit from hearing others' thinking, while students who are successful can have their thinking reinforced, whether by hearing another's similar explanation affirmed or having their own explanation affirmed.

Whether you include only one or all of these opportunities to use questions in these ways depends, of course, upon you, your students, and your lesson.

A Note on a Common Pitfall: Using Questions to Jump to Instruction

While questions can serve as a useful instructional tool in some situations, be wary of a possible misuse of questions when you are seeking to learn more about students' thinking. In our eagerness as teachers to help our students feel successful, we sometimes use questions as an instructional tool when we are intending instead to use questions to learn about how the student is thinking. When this happens, the questions can start to sound more directed, as they lead students to a correct answer. Here's an example:

Mrs. Meyers looks over Josh's shoulder and sees the following work on his paper.

Find six numbers that have a mean of 8.							
10	7	8	5	3	9	**42 ÷ 6**	7
13	7	11	5	3	12	**51**	8.5

Mrs. Meyers thinks: *(He's just using guess and check. I need to get him to see a better way to do this.)*	*Mrs. Meyers says,* *"So what's another way you could find six numbers with a mean of 8? How do you find a mean?"*
Josh thinks: *(I don't know how to do this. I hate this.)*	*Josh mutters, "Add 'em up and divide by how many you have."*
Mrs. Meyers thinks: *(Great! I'll help him build from that to see a more efficient solution method.)*	*Mrs. Meyers says,* *"So what would you have started with, if you had divided something by 6 and ended up with 8?"*
Josh thinks: *(. . . What?)*	*Josh starts to erase his work. Mrs. Meyers says,*

	"You don't need to erase what you've already written. Just think about it; you have some number, you divide it by 6 and you end up with 8. What did you start with?"
Josh thinks: *(8 × 6 . . . 8 × 6 . . .)*	*Josh says,* *"48?"*
Mrs. Meyers thinks: *(Almost there; let's see if he can put it together.)*	*Mrs. Meyers says,* *"Right! So how could you find 6 numbers that add up to 48?"*

If questions are used almost exclusively to try to lead students to correct answers, they can also bring about negative and unintended effects. For example, if the student hears the implicit message, "When I ask you a question, I already have the right answer in mind, and it's your job to figure out how I want you to answer this question," the student can quickly decide that "questions from the teacher just mean I'm wrong or that I need to figure out the right answer for her question." For many students, this can quickly shut down their willingness to respond to questions and create an obstacle to implementing formative assessment effectively that can take a long time to undo. (See Chapter 7, "Establishing a Classroom Environment," for more information about encouraging a culture in which students feel safe providing their thinking.)

As a contrast, consider this set of questions that maintains a focus on understanding the student's thinking:

Mrs. Meyers thinks: *(He looks like he's using guess and check. He appears to be adjusting his guesses to try to home in on a mean of 8. I wonder if he has any method to his guessing.)*	*Mrs. Meyers says,* *"Let's see what you did here. Can you explain to me what you've been trying so far?"*	*Josh replies,* *"I tried 6 numbers and added them up, and they equaled 42. But that's too small, because I divided by 6, and the mean is 7. It needs to be 8, that's 1 higher, so I tried adding 3 to some of the numbers, but that was too big. So I need to try another set of numbers and make them smaller."*
Mrs. Meyers thinks: *(He's got the idea that he could increase the mean by increasing some of the numbers, but it looks like he's just increasing every other number. Check this with him.)*	*Mrs. Meyers says,* *"So how are you deciding which numbers to change? And how do you decide how much bigger to make them?"*	*Josh says,* *"I picked the three biggest numbers in the list and made them go up by 3 each."*
Mrs. Meyers thinks: *(Oh! Not what I thought.)*	*Mrs. Meyers says,* *"And why by 3?"*	*Josh says,* *"Because I needed the total to increase, but not by a whole lot, so I made it be 9 and spread it out over the three numbers. Wait." (He pauses, then writes 48 under 51, then writes 48 ÷ 6 = 8.) "It needs to add up to 48."*

In this case, Mrs. Meyers is intentionally holding off a few moments on providing direction or correction in order to learn more about Josh's thinking so she can better support him. She is ensuring that her interpretation of the evidence, the next part of using evidence that we discuss, is accurate. As is sometimes the case, posing questions about Josh's thinking gave him enough scaffolding to think through the problem and arrive at a solution himself.

Interpreting Evidence of Meeting the Success Criteria

Another part of using evidence in your classroom is interpreting the evidence, which involves answering the question, "What does this information (provided by the evidence) tell me about how my students are making progress toward the learning intention?" as shown in Figure 3.6.

Figure 3.6   Interpreting Evidence Component of Using Evidence

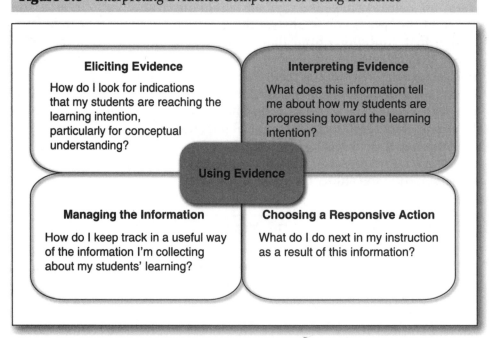

The following recommendations provide some guidance for your efforts in interpreting the evidence you gather.

> **Recommendation 5:** Focus your interpretation efforts on evidence related to how students are progressing toward the success criteria, and pay particular attention to evidence of meeting or partially meeting the criteria.

The emphasis in this recommendation is on the phrase *toward the success criteria*. Keeping your interpretation efforts focused specifically on the success criteria helps you maintain the same focus described by the learning intention. As you consider all the evidence you get from students in any given lesson, you may notice many things—errors, on one hand, and astute observations, on the other—that are not really pertinent to the success criteria or the learning intention. While you do not need to ignore them, they may need to sit more in the background for you as you focus on the matters related to the success criteria, so that you can keep students moving forward on the specific learning intention you've outlined for the lesson.

Striking a proper balance between attention to both what students understand as well as what they do not understand is essential to both gathering and interpreting evidence. Considering what is present in the students' thinking will enable you to make decisions about how to best structure experiences that will enable the students to begin to close the gap between where their understanding is and where you would like it to be.

For example, imagine a lesson for which the learning intention and success criteria are as follows:

LEARNING INTENTION AND SUCCESS CRITERIA

Learning Intention: Today we will learn how rhombuses and squares are related to each other.

Success Criteria:

1. I can identify whether a shape is a rhombus, a square, or neither.

2. I can explain how a rhombus is related to a square.

In the lesson, students are given different examples and nonexamples of rhombuses, and they are asked to decide if the statement "A rhombus is a square" is always, sometimes, or never true. They must justify their choices. Here are two responses from students:

A rhombus is always square because it has 4 equal sides and four 90-degree angles, but it is tilted sideways.

Student 1's Response

A rhombus is never a square because they are two different types of shapes. They are related because they both are quadrilaterals.

Student 2's Response

In this case, neither student has met the criteria completely, but both understand *something* about common characteristics between squares and rhombuses that can be expanded upon. Student 1 knows that rhombuses must have equal sides but mistakenly believes that the angles must also be right angles. The student also seems to believe that orientation of a shape is important to its classification. On the other hand, Student 2 recognizes that rhombuses, like squares, must have four sides, but it's not clear how much more the student knows. If the teacher's stance were to focus on whether the students "got" or "didn't get" the success criteria, then both students "didn't get it," and the teacher might choose to reteach the lesson or review the material. However, if the teacher's stance is to focus on the *extent* to which the students have met the success criteria and *what sense* the students are making out of the ideas, then both students show evidence of some understanding that the teacher can build from. Instruction therefore can more effectively target the places where students' understanding is lacking, and less time is spent reviewing the parts of the material that students do understand.

Recommendation 6: Interpret evidence at a whole-class level as well as an individual student level.

All teachers are faced daily with the dilemma of making sense of what their students know, both individually and collectively. With mathematics in particular, the knowledge is frequently developed by building new ideas on previous ones, and there is an emphasis on being able to solve a variety of kinds of problems. Because of this, it is perhaps easy to get caught up in paying attention more to what individual students need, because each student can have such different needs in terms of the mathematics. However, this recommendation calls for stepping back to look at the larger picture of your class and to remember to interpret evidence at a whole-class level as well.

When you gather evidence from a significant number of students, especially when it's from the entire class, one quick way to interpret the evidence is to look for patterns in what students *do* understand as well as what they don't. A pattern might be found across several responses from a single student, or you may notice clusters of students having similar issues or sharing similar understanding. In either case, identifying the patterns can help inform your instructional decisions about where to go next.

For example, in a lesson on proportional relationships, a teacher had the following learning intention and success criteria:

LEARNING INTENTION AND SUCCESS CRITERIA

Learning Intention: What does it mean for two sets of numbers in a table to be proportional to each other?

Success Criteria:

1. I can find a pair of numbers that are proportional to another pair.
2. I can use words, graphs, or other pictures to show whether two sets of numbers in a table are proportional to each other.

Mathematics Background: Proportional Relationships in Tables and Graphs

Two sets of numbers are proportional if they remain in the same ratio for any corresponding pair of values in the relationship. For example, if a student gets *paid $8 an hour for babysitting, then the amount of pay is proportional to the number of hours worked; the ratio of pay to hours is always 8:1.*

Hours	2	4	6	10	20
Pay	$16	$32	$48	$80	$160

If you graph this relationship, it appears as a straight line that passes through the point (0,0) on the graph. This is true of all proportional relationships—they are multiplicative, of the form y = kx. In this case, x is the number of hours, y is the pay, and k = 8.

Notice in the table above that both rows increase by a constant amount (though the amount is different for each row). Not all sets of values that increase by a constant amount are proportional. For example, when I rent a bicycle for sightseeing, I have to pay a $10 flat fee plus $7 per hour. The values in the following ratio table go up by constant amounts—hours increase by 1, and the fees increase by $7.

Hours	1	2	3	4	5
Rental Fee	$17	$24	$31	$38	$45

However, the ratio between each pair of hours and rental fees does not remain the same! Two hours to $24 is a ratio of 1:12, but 5 hours to $45 is a ratio of 1:9. If you graph this relationship, it also appears as a straight line, but it does not pass through the (0,0) on the graph because of the initial $10 flat fee. This relationship is not proportional, and the sets of numbers in the table are not proportional.

The teacher noticed several ways in which different students were having difficulty meeting the success criteria, namely,

- some students struggled with finding proportional pairs, using additive relationships rather than multiplicative relationships;
- some students were unsure how to show whether two sets of numbers in a table were proportional; and
- other students, who understood that proportions were made up of equivalent ratios, still were confused about how to identify proportional relationships in tables.

Focusing on all these different areas of difficulty could easily lead the teacher to feeling like he or she just needed to reteach the entire lesson in order to address all the various difficulties that were arising. However, let's add in additional information from the teacher's interpretation of the evidence about what students did understand:

- Many of the students understood that proportional relationships increase or decrease by a constant amount.
- All the students knew how to plot points on a graph.

Focusing on this evidence of understanding that the whole class—or at least a majority of the class—had, combined with looking for the strengths in students' understanding, led the teacher to structure the next day's lesson using graphs to compare proportional and nonproportional relationships, focusing on those that increase or decrease by a constant amount. The interpretation of evidence is closely linked to choosing a responsive action, the next part of using evidence.

Choosing a Responsive Action

Another part of using evidence is choosing a responsive action, which involves answering the question, "What do I do next in my instruction as a result of this information?" as shown in Figure 3.7.

Figure 3.7   Choosing a Responsive Action Component of Using Evidence

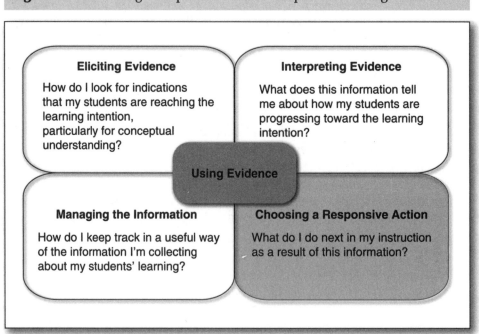

Eliciting Evidence

How do I look for indications that my students are reaching the learning intention, particularly for conceptual understanding?

Interpreting Evidence

What does this information tell me about how my students are progressing toward the learning intention?

Using Evidence

Managing the Information

How do I keep track in a useful way of the information I'm collecting about my students' learning?

Choosing a Responsive Action

What do I do next in my instruction as a result of this information?

The most important part of formative assessment comes hand in hand with interpreting the evidence you've collected: determining the appropriate next steps for students based on all the information you've collected. While it is one of the most important parts of formative assessment, research also shows that it is also very challenging for teachers (Heritage, Kim, Vendlinksi, & Herman, 2009).

In the preceding example, the teacher decided to use her students' strengths to try to develop deeper understanding about proportional relationships. In the next day's lesson, the class explored line graphs of different relationships, some of which were proportional and some of which were not, to determine another way to distinguish between the two. For some students, the graph provided a visual tool for finding equivalent ratios and using those to determine whether two sets of numbers were proportional. Providing further instruction, particularly further instruction specifically targeted to students' strengths and difficulties as a result of the evidence gathered, is just one of what we call possible *responsive actions* you can take. The next recommendation addresses these actions.

> **Recommendation 7:** Use interpretation of evidence against the success criteria to determine an appropriate responsive action, such as the following:
>
> - Gather more evidence.
> - Provide further instruction.
> - Provide formative feedback.
> - Move on.

At any time in a lesson when a teacher elicits and interprets evidence, there are a number of possible responsive actions, whether for an individual or for the whole class, and for formative assessment purposes, we focus on the four possibilities listed above. We tease them apart separately here to name them and discuss them, although in the reality of day-to-day work in the classroom, these responsive actions may be difficult to separate so discretely.

- **Gather more evidence.** In many cases, when a student gives you a response, the information you just gathered may be inconclusive. You may be scratching your head, feeling like you're still very unclear what the student is thinking. You might have gained *some* insight into student thinking but not enough for you to determine which of the other responsive actions is most appropriate. In this case, you simply need to gather more evidence. A follow-up open question, for example, is often effective.
- **Provide formative feedback.** When students show enough understanding or skill that some slight adjustments will allow them to meet the success criteria, then they may gain a lot from formative feedback. (We discuss this particular responsive action in detail in Chapter 4, "Providing and Using Formative Feedback.")
- **Provide further instruction.** If the evidence suggests that there is a significant gap between a student's current learning status and the success criteria, the student probably would not be able to make progress after receiving formative feedback. This may be because there is a significant barrier or misconception or because there are enough smaller issues that they collectively prevent the student from moving on. In such a case, taking a step back to address the issues through further instruction, such as approaching the content in a different way or addressing some prerequisite understandings (see Chapter 6, "Using Mathematics Learning Progressions"), is often an appropriate responsive action.
- **Move on.** Most often, this is appropriate when your students are on track with where you expected them to be, with regard to meeting the success criteria. This may be after a task has been completed and you check in to see where students are with a particular success criterion; even if they haven't met the criterion completely, they might be far enough along that they are ready to continue to the next task in your lesson plan. This also may be at the end of a lesson, when you review the evidence of the entire lesson against the whole set of success criteria. Perhaps they have met the success criteria and are

ready to continue; or, your students may be on the verge of understanding, and you know that one of the next lessons will help them solidify the concept.

Your choice of one of these next steps results from the way in which you think through the decision, which might be portrayed in Figure 3.8.

Figure 3.8 Deciding on a Responsive Action

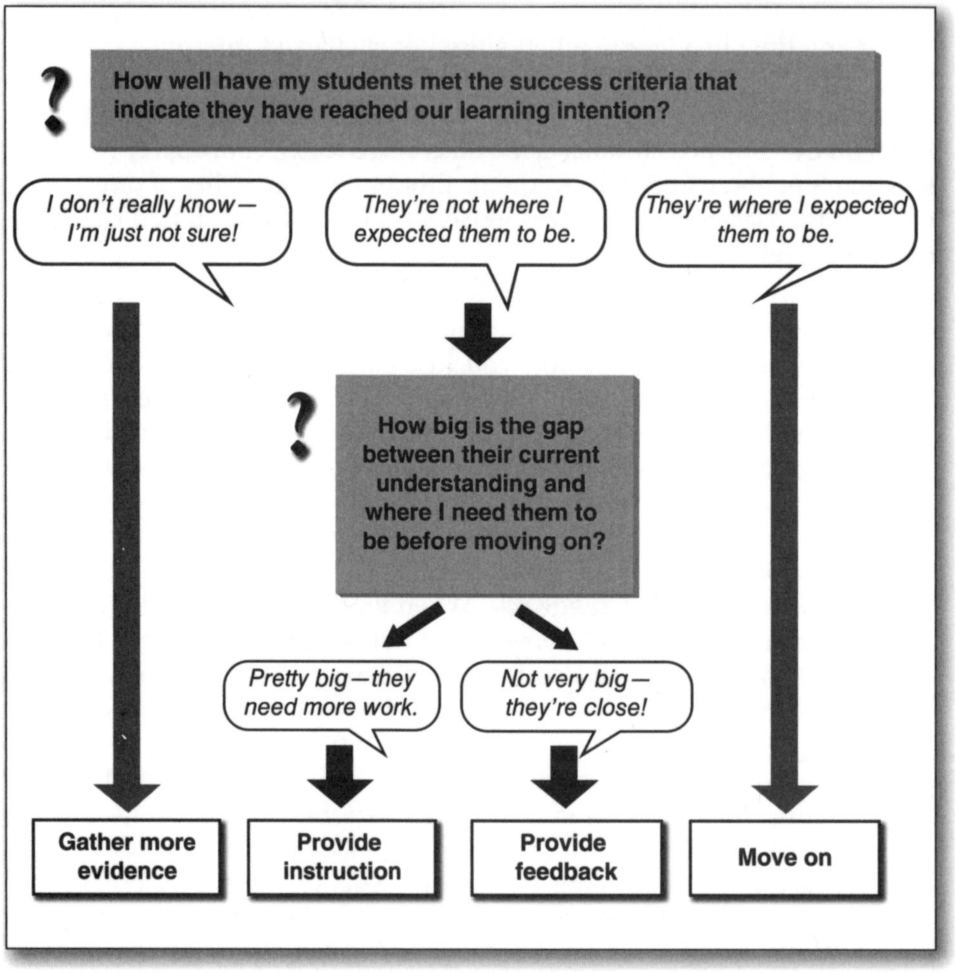

? How well have my students met the success criteria that indicate they have reached our learning intention?

I don't really know— I'm just not sure!

They're not where I expected them to be.

They're where I expected them to be.

? How big is the gap between their current understanding and where I need them to be before moving on?

Pretty big—they need more work.

Not very big— they're close!

Gather more evidence

Provide instruction

Provide feedback

Move on

Notice that although the flowchart suggests this is done for the class as a whole, the same decisions can be made for individual students. During the lesson, you might not expect students to have fully met any success criteria after an instructional activity—another activity might be planned to help them meet one or more criteria. Interpreting the evidence against the success criteria will still help you decide whether your students are where you expected, or need, them to be before moving on to the next instructional activity.

Choosing among these responsive actions can happen *during the lesson* or *after the lesson* is done. There are two main differences in how you think about responsive actions in these two instances. One is in the *immediacy* of your response and how rapidly you might employ multiple responsive actions; the other is the *amount and source* of your evidence. Let's look at each of these instances.

Responsive Actions During the Lesson

In the Formative Assessment Cycle, we are including two responsive action decision points using the starburst shape shown below.

Figure 3.9 Responsive Action Decision Points in the Formative Assessment Cycle

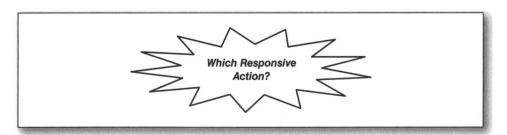

Figure 3.10 shows our evolving Formative Assessment Cycle with our first responsive action decision point, during the lesson. Notice that we have split Instructional Activities into two types: New Instruction and Further Instruction. By this, we mean *further instruction* to be activities that focus on the same content that students have already been working on, just providing different experiences with that content. *New instruction* includes activities that provide experiences with new (though similar) content. Notice also that instead of a single arrow moving to the feedback circle, there are four arrows. Trace the arrows and connect each of their destinations with one of the four responsive actions.

We are also now including a second question for the During the Lesson section: "How can learning move forward?" The ❷ by the responsive action decision point shows that the choice of responsive action is intended to answer that question.

Example: In Ms. Jenkins's lesson about connecting numerical patterns to visual patterns, she gathers evidence during the warm-up by looking at students' responses, and she interprets it to conclude that most students were able to find the pattern, so she feels confident they could move on to the planned Exploration Tasks.

From their work on Exploration Task 1, she realizes that most of her students provide reasonably clear descriptions of the patterns, meeting the first two success criteria (I can describe how a pattern of numbers continues, and I can describe how a corresponding visual pattern continues). However, two students are still unable to articulate the patterns or how they found them, and a few are able to describe their patterns but only with a struggle and using examples rather than direct explanations. She decides to pull out those having difficulty for further instruction, while the rest of class moves on with Exploration 2 (new instruction) as planned.

Figure 3.10 "During the Lesson" Responsive Actions in the Evolving Formative Assessment Cycle

BEFORE the lesson
Where are the students going?

Teacher determines the learning intention and the **success criteria** for the lesson

Teacher explains the **learning intention** and the **success criteria** for the lesson

Students understand the **learning intention** and success criteria for the lesson

DURING the lesson
1 *Where are the students now?*
2 *How can learning move forward?*

New Instruction

Teacher **elicits evidence** of student thinking

Students **respond to** elicitation task

1

Teacher **interprets evidence** against the success criteria

Students **self-evaluate** and/or **analyze peer work** against the success criteria

Which Responsive Action?

2

Formative Feedback
using success criteria

Further Instruction

AFTER the lesson
Where are the students going next?

Teacher determines the **learning intention** and the **success criteria** for the lesson

94

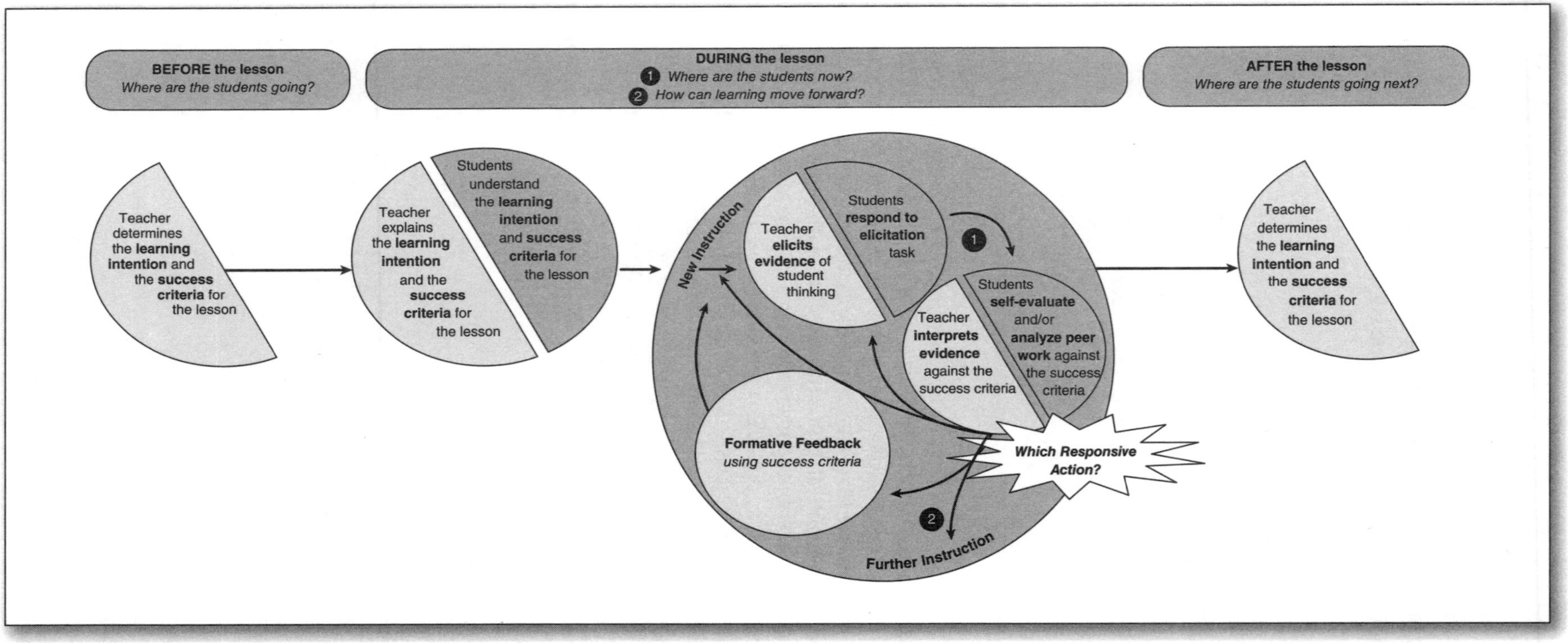

During the lesson, you have many opportunities to provide multiple responsive actions. The hinge-point questions we discussed with Recommendation 1 are one clear opportunity in which you will want to consider what responsive action to take, but nearly every interaction you have with students presents a potential for you to choose one of these actions. Perhaps more commonly during an interaction with students, you use a combination of different responsive actions with a student in real time to try to understand and respond to what the student needs, and as a result, your interpretation of the evidence you're getting will build and evolve as you do so:

- You might pose a question to see what a student is thinking (elicit evidence) and not be able to make sense of how the student answers (attempting to interpret the evidence).
- So you might then pose more questions (gather more evidence) until you can see (by interpreting the evidence) where the student is struggling.
- You could then pose something for the student to think about (provide formative feedback).
- At this point, you might leave the student alone to work through the feedback, returning a few minutes later to elicit evidence again. If you find the feedback was helpful, you might choose to embellish on the feedback by providing a bit more of a hint or cue for next steps, you might simply give the student more time and practice to solidify their understanding, or you might decide that the student is ready for the next activity (move on). If the feedback was not helpful, you then might try explaining something to the student (provide further instruction—in this case, in the moment and just to this particular student) and see whether the explanation was helpful.
- The results, and whether you found the issue to be a common one, will determine whether you decide to keep working with this student, to step back and address an issue with the whole class, or to note that this student needs particular one-on-one help as the rest of the class moves forward.

The patterns mentioned in Recommendation 6 could be used to assign students to small groups for a next activity. In the proportional reasoning example provided with Recommendation 6, the teacher might decide to group students in a follow-up activity so that each group had at least one "expert" student with regard to each of several different tools for thinking about proportional relationships (such as using ratio tables or using a graph), to help them guide their work and explain the difficulty to those who need help. Alternatively, the teacher might group students with similar approaches or difficulties so that he or she can tailor instructional activities to best suit where each group of students is with respect to the success criteria.

As you compare student responses to the success criteria, for example by looking at whiteboard responses, a quick separation into *met*, *partially met*, and *not met* can provide some easy insight into how close the class is, collectively, to reaching the learning intention and where you (and they) need to focus your energies to help them achieve what you intended. An important caveat here is to revisit student needs often and regroup accordingly, since students who need extra help with one concept or skill may have solid understanding of another concept.

Learning Resource: Go to Resources.Corwin.com/ CreightonMathFormativeAssessment to access this resource.

Responsive Actions—During. This interactive web page provides some practice deciding which responsive action is reasonable for individual students, small groups, or a whole class.

Reference Resource: Go to Resources.Corwin.com/ CreightonMathFormativeAssessment to access this resource.

Chapter 3 Summary Cards. This printable file provides Summary Cards that include the four responsive actions.

Responsive Actions After the Lesson

The same four responsive actions apply here to your decision making, as you determine how to structure the next day. Figure 3.11 shows the next stage of our evolving Formative Assessment Cycle, with an After the Lesson responsive action decision point included. Take a moment to notice the arrows emerging from this decision point. *Moving on* involves developing a new, subsequent learning intention and set of success criteria, and so it appears in the Formative Assessment Cycle as the "new LI, new lesson" option. The other three choices (Gather More Evidence, Provide Formative Feedback, and Provide Further Instruction) are embedded within the choice to adjust instruction with either a revised learning intention and success criteria or with the same learning intention and success criteria but a different accompanying lesson. The following paragraphs explain those choices.

Choosing to gather more evidence: For example, as you reflect on the lesson and the evidence you gathered, you might decide that you are still unsure where your students are. It may be that some issues still remain, since you're not getting the clear information you expected, or it may be that your elicitation strategies didn't quite get at that information. (See page 73 for a note about misaligning the elicitation task with the evidence sought.) Either way, you want to try again, so you might decide to stay with the *same learning intention and success criteria* as you gather more evidence related to them. Or, you might decide that the success criteria were partially met, and so you can *revise your learning intention and success criteria* to focus on the part that isn't clear yet. See Figure 3.12.

Choosing to provide formative feedback: If you have decided that the appropriate responsive action is to provide formative feedback to help students with the part of the success criteria not yet met, then you may want to start the class with that feedback and an activity to allow

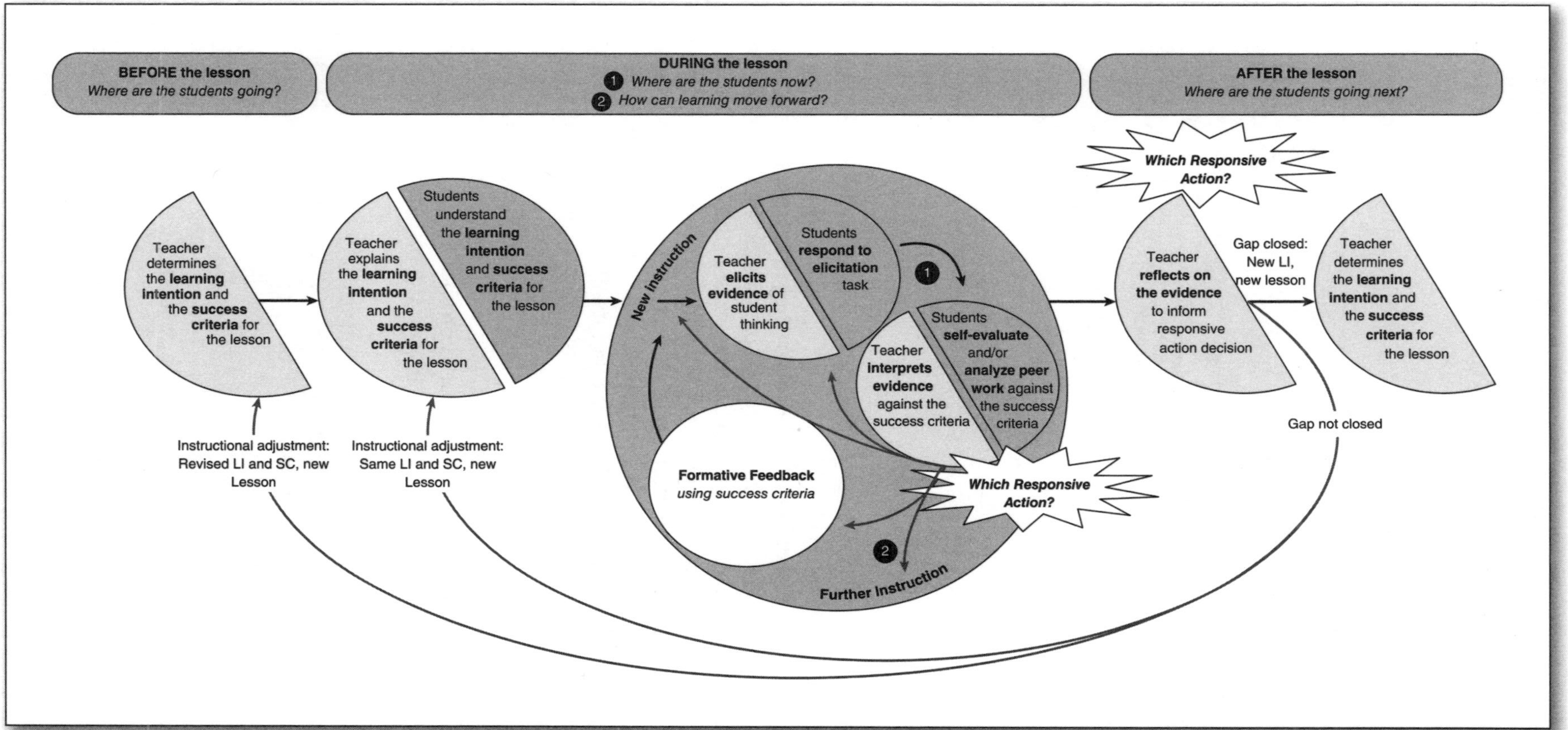

Figure 3.11 After the Lesson Responsive Actions in the Evolving Formative Assessment Cycle

them to respond to it. Again, you might decide to stay with the *same learning intention and success criteria*, or you might decide to *revise your learning intention and success criteria* to focus students' attention and efforts.

Figure 3.12   Returning to Choosing Learning Intentions and Success Criteria

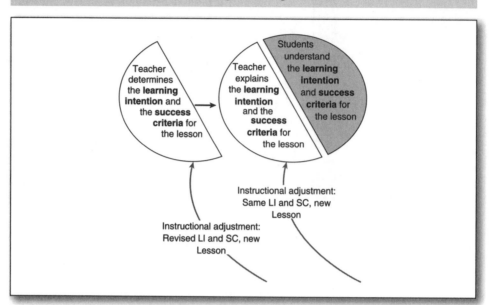

Choosing to provide further instruction: As a final option, you might feel that students have not progressed enough that feedback will be sufficient. Perhaps the activities simply didn't have the impact you intended, or they uncovered gaps in more foundational concepts; perhaps your learning intention presented a larger leap in understanding than you anticipated, or perhaps the students have a gap in their understanding that needs to be addressed before they can understand the new content. Whatever the reason, if providing further instruction is needed, you have the decision of whether you can stay with the *same learning intention and success criteria* (for example, so you can address the content from a different perspective) or whether you need to *revise your learning intention and success criteria* (for example, to provide needed background information or to take a smaller step in building the concepts involved).

> *Example:* At the end of class, Ms. Jenkins returns her students' attention to the success criteria, and she projects the exit ticket prompts using the document camera:
>
> 1. In the pattern we just examined, describe as completely as you can how the number pattern is related to the visual pattern.
>
> 2. From today's lesson, what success criteria are still presenting a challenge for you? Explain your reasons for your answer.

She asks, "Which of our success criteria, 1, 2, or 3, is best addressed by this first prompt question? Hold up the number of fingers." After a moment, most students have three fingers up. She confirms the choice and gently reminds them that Success Criteria 1 and 2 are about determining and describing the patterns, but only Criterion 3 is about making that final connection between them. As she passed out slips of paper for them to write their responses, she adds, "For Question 2, please think about all three success criteria when you write your answer. Be as specific as you can!"

As Ms. Jenkins plans for the next day's class, she reviews what she saw during the class as well as the student exit tickets. She determines that the students basically fall into three groups in terms of their understanding of how the number patterns relate to the visual patterns: (1) those who cannot yet describe the growth of either the visual or numeric pattern, (2) those can describe the growth of the visual pattern and the numeric pattern but can't relate them to each other, and (3) those who can describe the connection reasonably well and are ready for something a little more challenging. The first two groups held enough students that she decides not to move to generalizing patterns but rather to spend one more day on relating numeric and visual patterns. She plans to provide some differentiated instruction using the three groups she already identified.

 Planning Resource: Go to Resources.Corwin.com/ CreightonMathFormativeAssessment to access this resource.

 Using Exit Tickets as Evidence. This printable file describes a process to focus examining exit tickets for trends in student responses, rather than focusing on individual responses.

 Learning Resource: Go to Resources.Corwin.com/ CreightonMathFormativeAssessment to access this resource.

 Responsive Actions—After. This interactive web page provides some practice deciding which responsive action is reasonable after the lesson is concluded. The interactive also walks you through the Sorting Exit Tickets process (discussed earlier).

 Reference Resource: Go to Resources.Corwin.com/ CreightonMathFormativeAssessment to access this resource.

 Chapter 3 Summary Cards. This printable file provides Summary Cards that include the full process for eliciting and interpreting evidence.

Managing the Information

The remaining part of using evidence is managing the information, which involves answering the question, "How do I keep track in a useful way of the information I'm collecting about my students' learning?" (see Figure 3.13). The focus of our recommendation for this part of using evidence is on keeping it manageable.

Figure 3.13   Managing the Information Component of Using Evidence

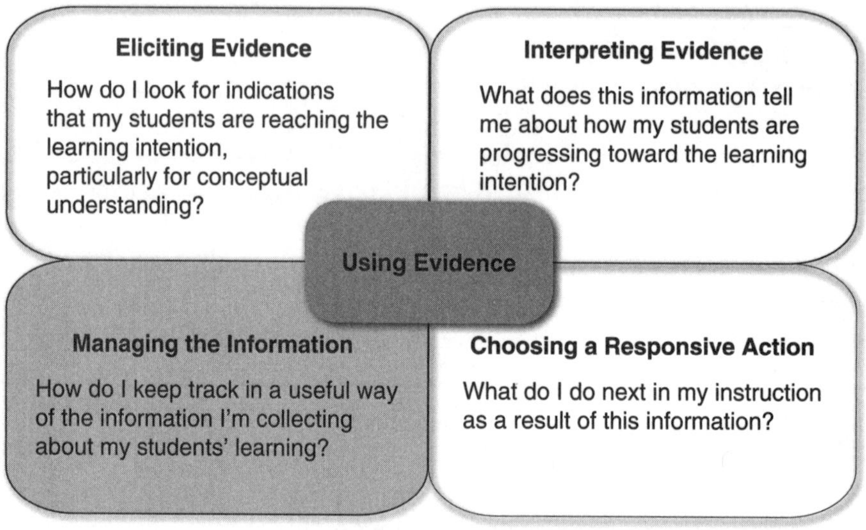

Eliciting Evidence

How do I look for indications that my students are reaching the learning intention, particularly for conceptual understanding?

Interpreting Evidence

What does this information tell me about how my students are progressing toward the learning intention?

Using Evidence

Managing the Information

How do I keep track in a useful way of the information I'm collecting about my students' learning?

Choosing a Responsive Action

What do I do next in my instruction as a result of this information?

Recommendation 8: Be selective about what information you need to record, and keep it simple.

Although many teachers already have strategies for collecting evidence and do this regularly, doing so as part of formative assessment might seem like a lot of information to manage. This need not be the case. The difference between eliciting evidence for formative assessment purposes and what most teachers already do lies in focusing your evidence-gathering efforts specifically on—and primarily on—the success criteria. Keeping that in mind can help you prioritize which information you pay attention to.

What's more, much of the information gathered is fleeting—once you gather the information, interpret it, and react to it in the moment, it has served its purpose and does not need to be recorded for future reference in the same way that summative assessment information might need to be.

Let's look again at Mrs. Keller's lesson about slope and rate of change, to consider the ways in which she used many of the recommendations for eliciting evidence. There were several points at which she elicited evidence from her students:

- At the start of class, she had a brief whole-class discussion on what they thought *rate of change* meant.
- She created a class list of the ideas that the class revised and refined later in the lesson.
- As students worked on identifying rates of change in problems, she circulated and listened in on the discussions they were having.
- She revisited students' understanding of rate of change in a follow-up discussion.
- She provided an exit ticket prompt to get students to summarize their current understanding.

As was true for Mrs. Keller, much of your evidence gathering can occur organically as part of the lesson, and not all of it needs to be recorded. Mrs. Keller could have chosen to use a camera, a mobile phone, or a tablet device to snap a before and after picture of the class list of ideas about rate of change. She also has the exit tickets as concrete evidence she can save and refer to. Some of her evidence is meant to be more fleeting, simply giving her in-the-moment information from which she can make in-the-moment instructional decisions about where to go next with the lesson. If there is information that you decide you want to record during the lesson, perhaps while you are circulating and checking in on students, creating a simple recording sheet with students' names and one or all of the success criteria on it might be sufficient for making some quick notes. Or you may choose to create a sheet of labels, with a student name on each one; as you notice something noteworthy, you can note it on the student's label; then after class, peel off the label and add it to a notebook page devoted to that student. Over time, the notes from various labels can help paint a picture of an individual student's learning.

TEACHERS' VOICES

We each carry a ring of the students' names on small colored index cards on our lanyard. When a student is doing something that will help him or her to move forward, says something mathematically interesting or important, agrees or disagrees with an idea and provides evidence, the student gets a "+" on him or her card. At the end of the grading period, I look at the cards and use them to help determine a participation grade. The grade is typically less than 5% of the overall grade, but it serves as an incredible motivational tool.

The important thing here is to be selective about what information you want to record and to think about that when planning your lesson. It can be tempting to record lots of information about your students, but don't confuse *formative* assessment purposes with *summative* assessment purposes. Keep your formative assessment information recording informative but manageable.

■ THE STUDENT'S ROLE AND HOW YOU CAN DEVELOP AND SUPPORT IT

We say a few words here about your students' role and how you can support it, and we provide a more in-depth discussion of the student role in Chapter 5, "Developing Student Ownership and Involvement in Your Students." Students have a very important and active role to play with the use of evidence. Part of their role is to participate in the class activities, providing honest answers to questions so that you, and they, can accurately uncover how they're thinking about mathematics. Another important role is *self-monitoring*, which is one aspect of self-regulation, and students need to understand the role that evidence can play in it. While it's important for them to consider whether they *believe* they can meet a success criterion, they also need to understand how to verify that belief.

Participation

When you elicit evidence from your students, you are asking them to communicate something to you about their thinking or their understanding. Initially, they may be more or less comfortable telling you what answer they got or even what they did to get their answer. But the kind of explanation of their thinking that you're seeking goes beyond those initial comments from students. If students have not done this a lot, they need time and experience, and in some cases explicit instruction, to build their communication skills.

For example, Ms. Genovese talked to her students at the start of the school year about her desire to understand their thinking, not just to evaluate their answers. She also talked about some of the things she would be doing with them in class to try to better understand their thinking, including asking them lots of questions, even when their responses were correct; asking them to do lots of explaining aloud to others in class; and making sure that everyone was letting her know how well they felt they understood something. She comes back to this throughout the year and reminds students of these points, as she does in this interaction:

Ms. G: *OK, Eric, can you tell us how you figured out 45% of 20?*

Eric: *Um, I knew that 50% of 20 was 10. But it asked for 45%, so it's 9.*

Ms. G: *And how do you know it was 9?*

   (Eric's facial expression changes to worried, he pauses and just looks at Ms. Genovese, worried that her question means he is wrong.)

Ms. G: *(Reassuringly) Your answer is correct! I'd just like to hear what you were thinking to come up with that answer. (She addresses the whole class.) Remember at the start of the year, we talked about how I will often ask you questions whether or not your answer is correct, because I don't just want to know what answer you got but how you got it, how you were thinking. And when someone else is explaining their thinking, I'd like you to listen and decide whether that person is thinking in the same way that you did or differently than you did. So Eric, tell us how you landed on 9.*

Eric: *Well . . . 50% of 20 is 10. And I needed to go down 5%. But 10% is easier, so I did 10% of 20, which is 2 and took half of that to get 5%. So 5% was 1, and I did 10 – 1 and got 9.*

Ms. G: *Who else thought about it in the same way that Eric did? (Several students raise their hand.) OK, would any of you like to add to what Eric said?*

Certainly, students' willingness to participate and share their thinking depends very much on a classroom environment in which students feel safe in responding. (See Chapter 7, "Establishing a Classroom Environment.") They have some understanding of why their response is needed by the teacher and see that whether or not they respond does matter to the teacher. For example, Ms. Genovese likes to have her students use whiteboards to record and collectively hold up their answers for her to scan quickly; when she does this, she waits until *all* students have written something down and are holding up a response. At the beginning of the year, she often prompts students who aren't holding up their whiteboards, saying, "I would like everyone to share their responses on their whiteboards." In time, students learn that a nonresponse will be noticed, and so she uses the phrase less and less over the course of the year.

In a classroom conducive to formative assessment practices, a teacher may share his or her rationale for asking students many questions in order to convey to students that he or she is very interested in how the students are thinking and are making sense of the mathematics. Students need to know their honest responses are important, even if they're wrong or are showing that they're not confident in their abilities, so the teacher can make needed adjustments.

From the preceding discussion, you may have your own ideas about what you can do to support your students in fully participating in your evidence-gathering efforts. You may want to include the following in your efforts:

- Ask follow-up questions for both correct and incorrect answers—make "explaining your thinking" a regular part of responding in class.
- Ask students to offer different ideas about an approach to a problem.
- Be an attentive and curious listener when your students are talking. Often, students have a seed of correct thinking in an incorrect answer that may just be misapplied in the current situation.
- Listen for what your students *do* understand, and think about ways to connect to their existing understanding as you address their difficulties.

Self-Monitoring

The first two questions in Thinking Like a Self-Regulating Learner (Figure 3.14) focus on self-monitoring—identifying how well you are learning what you are meant to be learning.

In order for students to become more self-regulating, there are two key things for them to learn related to evidence gathering and interpretation

Figure 3.14 Thinking Like a Self-Regulating Learner

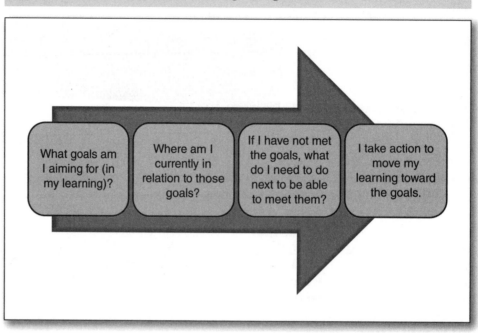

and self-monitoring: (1) learning to see their work as the evidence that they compare to the success criteria and (2) distinguishing between their actual mathematical work and how confident they feel about their mathematical work.

First, students need to learn that while you are evaluating their work against the success criteria, they too are responsible for that evaluation. You can model for them the kind of self-monitoring they can eventually do independently, by providing students with structures and regular opportunities to pause, revisiting the success criteria in the midst of a lesson, and thinking about how closely their work matches the success criteria. They may also need encouragement to see their work as a *work in progress*, not a completed task that deserves no further attention once it's done. When students understand that they can continue to refine their work by identifying where it meets the success criteria and what the work needs to better meet the success criteria, then they will begin to pay attention to the success criteria in a more intentional way.

Second, students need to learn to distinguish between their actual work and their confidence in their work, or even their reaction to whether they liked the lesson that day. In one sixth-grade classroom, students who were new to formative assessment practices were asked to indicate how well they met each of the three success criteria for the lesson. Many students marked all three success criteria positively, even though their exit tickets showed that they were not able to solidly complete two sample problems from the day. In their written comments, it was clear that some of them felt that they understood the concept although, in actuality, they did not yet understand it. Over time, the students needed reinforcement that they should be looking for *evidence* of meeting the success criteria.

Again, you may now have some ideas about what you can do to help your students learn to use their work as evidence to monitor their own learning. The following are helpful:

- Point out to students that you, and they, can return to the success criteria when you want to think about the quality of their work. Help students understand that their work is the evidence that they can compare to the success criteria.
- Help students learn to distinguish between looking at whether they liked the lesson activities, how confident they feel that they can do the work, and what they actually did in their work. While the first two are still valued, they do not provide evidence of success.

We return to supporting student self-regulation in Chapter 5, "Developing Student Ownership and Involvement in Your Students."

CONCLUSION ■

In this chapter, we focused on what constitutes *evidence* of students' understanding of a learning intention, and both the teacher's and students' roles in gathering and interpreting evidence. In concert with the learning intention and success criteria, the evidence provides a way to determine "Where am I currently in my learning, in relation to the learning goals?" We've also discussed how an informed use of evidence that is focused on the success criteria—both when gathering and interpreting—can help teachers make instructional decisions to more effectively target those aspects of students' understanding that move students' learning forward. Such a use of evidence also can model for students a process that they can learn to apply on their own to advance their own learning.

In the next chapter, "Providing and Using Formative Feedback," we take a closer look at one of the four possible responsive actions introduced in this chapter. Formative feedback is a critical element of implementing formative assessment practices, bringing together learning intentions, success criteria, and evidence to move students' learning forward in a substantive way. It helps students and teachers address both the third question in our self-regulation model, "If I have not met the goals, what do I need to do next to be able to meet them?" and the final step of the self-regulation model, "I am able to take action to move my learning toward the goals."

Recommendations in This Chapter

1. Create opportunities during the lesson to gather evidence related to each of the success criteria.

2. Consider from whom you need evidence. You might not need it from all students all the time on all parts of the lesson.

3. Elicit evidence about correct thinking and correct responses as often as you do about incorrect thinking and responses.

4. Use questions (at least some of the time!) to learn more about students' mathematical thinking.

5. Focus your interpretation efforts on evidence related to how students are progressing toward the success criteria, and pay particular attention to evidence of meeting or partially meeting the criteria.

6. Interpret evidence at a whole-class level as well as an individual student level.

7. Use interpretation of evidence against the success criteria to determine an appropriate responsive action, such as the following:
   • Gather more evidence.
   • Provide further instruction.
   • Provide formative feedback.
   • Move on.

8. Be selective about what information you need to record, and keep it simple.

Reference Resources: Go to Resources.Corwin.com/ CreightonMathFormativeAssessment to access these resources.

Chapter 3 Summary Cards. This printable file provides Summary Cards that include recommendations in this chapter.

Formative Assessment Recommendations. This printable file includes a summary of all the recommendations in this book.

■ RESOURCES

The following sections present some resources that can help you implement the recommendations. Each of these resources is referenced in earlier sections of the chapter, but here we provide a consolidated list. All resources can be found at Resources.Corwin.com/CreightonMath FormativeAssessment.

Learning Resources

These resources support your learning about formative assessment practices related to evidence gathering and interpreting:

• **Responsive Actions—During** provides practice deciding which responsive action is reasonable, for individual students, small groups, or a whole class.
• **Responsive Actions—After** provides practice deciding which responsive action is reasonable after the lesson is concluded.

Reference Resources

These resources summarize key ideas about formative assessment practices related to evidence gathering and interpreting:

- **Chapter 3 Summary Cards** provides index-sized Summary Cards for the following:

  o **Process for Eliciting and Interpreting Evidence** summarizes evidence related steps from planning for the lesson to enacting the lesson to determining next steps after the lesson.
  o **Considerations for Planning for Evidence Gathering** summarizes considerations when planning for eliciting evidence.
  o **Responsive Actions** summarizes four responsive actions.
  o **Recommendations for Eliciting and Interpreting Evidence** includes all the recommendations from this chapter.

- **Formative Assessment Recommendations** is a printable PDF that includes all the recommendations from this book.

Planning Resources

These planning guidelines and related tools are resources to support your lesson planning when integrating learning intentions and success criteria and evidence into your mathematics lessons:

- **Using Exit Tickets as Evidence** describes a way to sort exit tickets to inform your work for the next day.
- **Guiding Questions for Formative Assessment** includes guiding questions to help you plan for eliciting evidence for different purposes as well as examples of techniques suited for a particular purpose.
- **Formative Assessment Planner Templates** can be used for creating a formative assessment plan for a lesson. These templates will be revisited in later chapters, so for now pay particular attention to the portions related to setting learning intentions and success criteria and planning for how and when to collect evidence.

Classroom Resources

These resources illustrate various classroom routines that you can use during instruction when gathering and interpreting evidence with your students. Each routine provides a structure that you can use or adapt to routinize your practice around the use of evidence.

- **Framework for Using LI and SC With Students** describes a way to help students understand and use learning intentions (LI) and success criteria (SC) during a lesson. It describes actions for both teacher and students to take at the start of the lesson, at midway points during the lesson, and at the end of the lesson.
- **Revisiting Strategies** includes strategies for use when eliciting evidence related to LI and SC during the lesson. More about these strategies is included in Chapter 5.

- **Wrapping-Up Strategies** includes strategies for eliciting evidence related to LI and SC to wrap up a lesson. More about these strategies is included in Chapter 5.
- **Teacher Summary Cards for the Strategies** is a set of large-sized cards that provide a step-by-step summary of *teacher* moves for the strategies described earlier.
- **Elicitation Techniques** lists formative assessment classroom techniques (FACTS) for eliciting evidence that can be found in the book *Mathematics Formative Assessment: 75 Practical Strategies for Linking Assessment, Instruction, and Learning* by Keeley and Tobey (2011):
  - Commit and Toss
  - Fact First Questioning
  - Response Cards
  - Sticky Bars
  - Human Scatter Graph
  - I used to think . . . but now I think
  - Muddiest Point
  - POMS

- **Elicitation Strategies** includes a description of additional elicitation techniques not found in the FACTS resource:
  - Flip the question
  - Error Analysis

Classroom Materials

This resource can be copied or re-created to be used in your classroom:

- **Student Summary Cards for Strategies** is a set of large-sized cards that provide a step-by-step summary of *student* moves for the strategies described earlier. These cards can be distributed to each student or pairs of students for use when learning about a strategy.

4

Providing and Using Formative Feedback

In Mr. Carlton's eighth-grade algebra class, students are learning about setting up and solving equations, and today, students are working on creating an equation that fits a word problem, then solving it. He identified his learning intention and success criteria as the following:

LEARNING INTENTION AND SUCCESS CRITERIA

Learning Intention: We are learning today how we can use an equation to represent a word problem.

Success Criteria:

1. I can set up appropriate equations for word problems.

2. I can explain how the variables, numbers, and operations in the equations represent the word problems.

3. I can justify my equations by checking the reasonableness of my solutions.

The first problem of several that students are working on is included in the following box:

On e-Music, you can download an entire album for $12, and individual songs are $1.25 each. Jake has an e-Music gift card with $25 on it. If he buys one album, how many additional individual songs can he buy?

1. What should the variable represent in this problem?

2. Write an algebraic equation that could be used to help you solve this problem.

Mr. Carlton is looking for the equation $1.25x + 12 = 25$. He walks up to Emily's desk, and he sees the following work on her paper:

```
x = songs

  12x + 1.25 = 25
 -12          -12
 _____          ⎛           ⎞
     x + 1.25 = 13            ⎜  $ 11.75   ⎟
         -1.25 = -1.25        ⎝           ⎠
 _____
     x       = 11.75
```

While he notices one major mistake (subtracting 12 from both sides), he is also looking for things Emily has done correctly. Since the learning intention is focused on the relationship between a word problem and an equation, he starts by asking her about her thinking in setting up the equation.

Mr. Carlton: *Let's look at what you did here . . . OK. Tell me about how you set up your equation.*

Emily: *Well . . . you have to add up the different numbers to equal $25, and there are songs and an album so you need all the money for the songs and the money for the album.*

Mr. Carlton: *OK. It looks like you're thinking about adding up some amount for songs and some amount for album and adding them up to equal $25, so that part of your thinking is correct. You haven't quite translated your thinking into a correct equation yet, though; which of our success criteria is that?*

Emily: *(Looks up at the board) Um . . . the first one?*

Mr. Carlton: *Right. Maybe it would help to think about the cost of just 2 or 3 songs and the expression you would use to figure that out. Try that, and I'll check back with you in a few minutes.*

Mr. Carlton determined that Emily understood enough that she would benefit from feedback (one of the responsive actions we mentioned in Chapter 3, "Gathering, Interpreting, and Acting on Evidence"). His feedback was crafted in a particular way that makes it *formative feedback*. In this chapter, we look more at this particular responsive action, which can seem simple but takes practice and care to do well.

■ WHAT IS "FORMATIVE FEEDBACK"?

Formative feedback is feedback to a student that the student uses to move his or her learning forward. If a student has not yet reached the learning intention, formative feedback can be used to help the student adjust or revise his or her understanding and resulting work to move closer to reaching it. The information provided by the feedback should help the student know both what lines of reasoning or approaches to keep and what to rethink or revise in subsequent work. If a student *has* reached the learning intention, formative feedback helps confirm for the student not only *that* he or she has reached it but also what it is about his or her work that met the success criteria for the learning intention.

Since providing formative feedback is a responsive action, it appears in the Formative Assessment Cycle within the During the Lesson section as an option for "Which Responsive Action?" (See Figure 4.1.) For After the Lesson, formative feedback would be given as part of the next lesson or as a follow-up activity before the new lesson begins. This option is not explicit in the cycle diagram.

Example: *Ms. Jenkins's math class is just starting work on a prealgebra unit that includes learning about the relationship between visual patterns and numerical patterns, in preparation for studying generalized algebraic expressions. Her learning intention and success criteria for the lesson are*

Today, we will learn to relate growth in numeric patterns to growth in corresponding visual patterns.

1. *I can describe how a pattern of numbers continues.*
2. *I can describe how a corresponding visual pattern continues.*
3. *I can explain how the way a number pattern grows relates to the way the visual pattern grows.*

As students work on Exploration Task 1, finding and connecting a numerical pattern in a table to a visual pattern, Ms. Jenkins listens to the students' conversations and looks over their work. One pair appears to have found both patterns, so Ms. Jenkins asks about the connection between the patterns.

X	1	2	3	4
Y	3	5	7	?

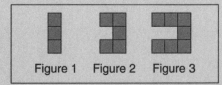

Figure 1   Figure 2   Figure 3

"I think I get it," Chase says. "The numbers in the table increase, and the ends on the pictures get longer." John nods and adds, "So the numbers go up by 2, and the pictures have 2 more squares added each time. Is that right?"

"Good work," Ms. Jenkins says. "You've described how each of the two patterns grow, so you're meeting the first two success criteria. The third criterion is about

(Continued)

Figure 4.1 The Evolving Formative Assessment Cycle (with feedback)

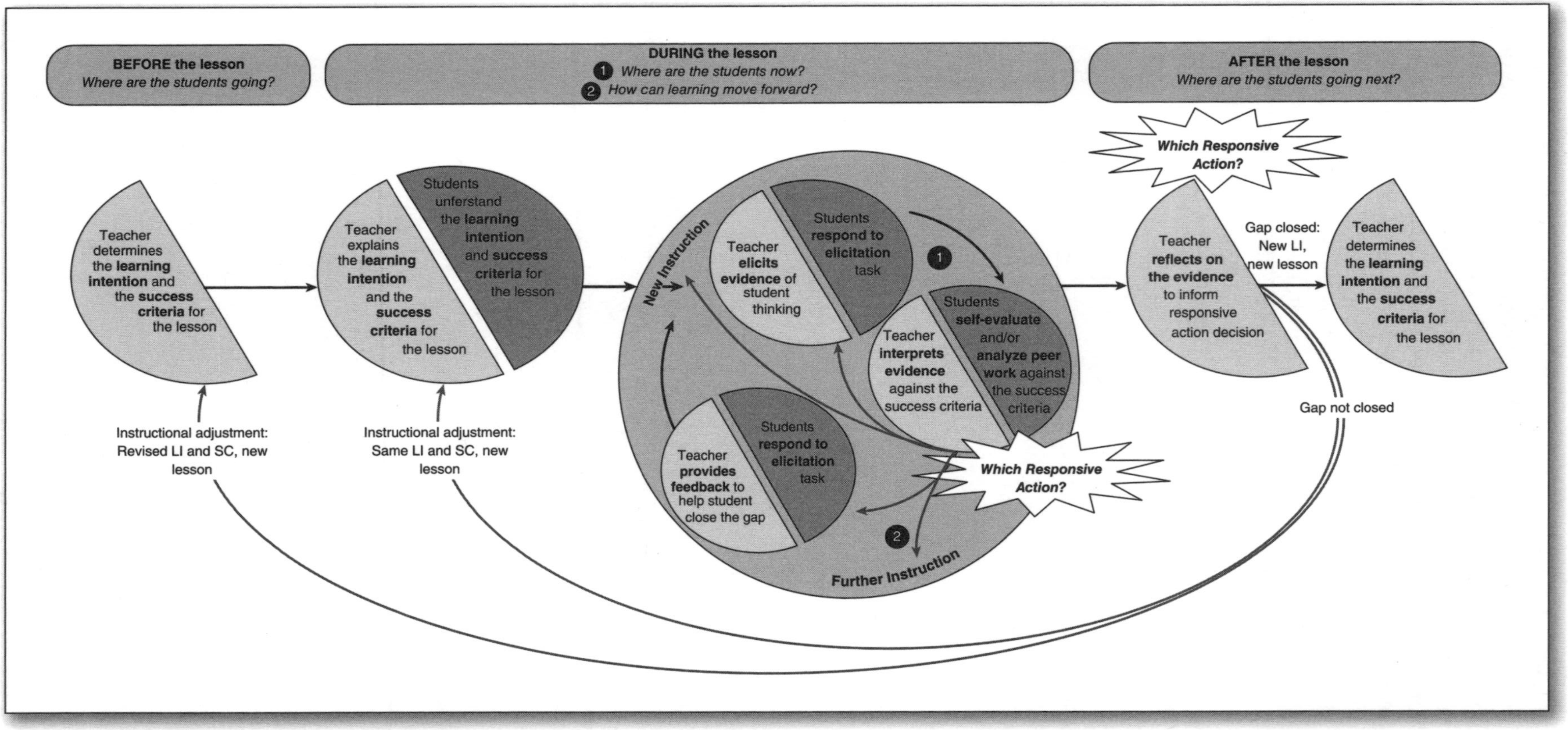

BEFORE the lesson
Where are the students going?

DURING the lesson
1 *Where are the students now?*
2 *How can learning move forward?*

AFTER the lesson
Where are the students going next?

Teacher determines the learning intention and the **success criteria** for the lesson

Instructional adjustment: Revised LI and SC, new lesson

Teacher explains the **learning intention** and the **success criteria** for the lesson

Students **understand** the learning intention and **success** criteria for the lesson

Instructional adjustment: Same LI and SC, new lesson

New Instruction

Teacher **elicits evidence** of student thinking

Students **respond to** elicitation task

1

Students **self-evaluate** and/or analyze peer work against the success criteria

Teacher **interprets evidence** against the success criteria

Which Responsive Action?

2

Further Instruction

Teacher **provides feedback** to help student close the gap

Students **respond to** elicitation task

Which Responsive Action?

Gap not closed

Teacher **reflects on the evidence** to inform responsive action decision

Gap closed: New LI, new lesson

Teacher determines the learning intention and the **success criteria** for the lesson

112

(Continued)

the connection between the patterns. John, you're being specific about how much each pattern is growing, and that's part of the connection, but I want you to be able to use the visual pattern to say why the numerical pattern would go up by 2. A pattern like '8, 10, 12' also grows by 2s, but '8, 10, 12' isn't part of this pattern—there's no connection." She writes 8, 10, and 12 as a new row in their table:

X	1	2	3	4
Y	3	5	7	?
	8	10	12	

"If you can tell me which parts of the table correspond to which parts of the figure, you can give me a more specific explanation about the connection between the original patterns that wouldn't apply to 8, 10, 12. I'll let you think about it and come back. Do you understand what you're supposed to talk about?"

"I think so," Chase said. *"We need to figure out how the number pattern matches up to the picture pattern."* Satisfied that this will get the pair closer to describing the connection, Ms. Jenkins moves on to see how the next pair is doing.

At the conclusion of a whole-class discussion after Exploration Task 1, Ms. Jenkins interprets the evidence from the students' work and the discussion. She provides formative feedback to the whole class by saying, "I heard some clear explanations of how the number patterns continue and how visual patterns continue—that shows me that, as a class, we're meeting the first two criteria. However, I'm not sure everyone is clear about our third success criterion, explaining how number pattern growth relates to visual pattern growth, so we're going to think a bit more on that as we try the next exploration. First, can I have a volunteer to tell the class what you think that success criterion means?"

For formative feedback to have the desired impact, giving students an opportunity to respond to the feedback is equally vital. This opportunity allows them to take action to move their learning closer to reaching the learning intention. As you'll see later in the chapter, if students stop short of taking action on the feedback they receive, the benefit of the feedback is reduced to it being merely additional information for the student to file away for another time, and it has much less impact on a student's learning than it could have.

Now let's consider how formative feedback fits into thinking like a self-regulating learner, shown in Figure 4.2.

Following the arrow from left to right, recall that learning intentions help students understand what goals they are aiming for in their learning. Both the corresponding success criteria and the evidence gathered and interpreted in relation to those success criteria help students understand where they are currently in relation to those goals. Formative feedback also helps students understand where they are currently in relation to those goals, as well as understand the vital next piece for moving their learning forward: If they have not yet met the goals, understanding what

Figure 4.2   Feedback in Self-Regulation

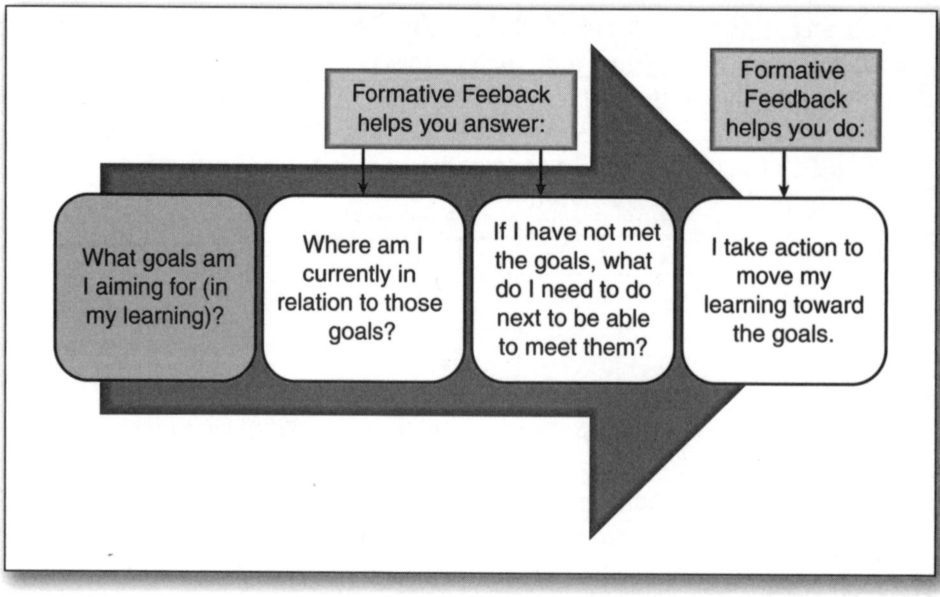

to do next in order to meet the learning goals. Look again at Mr. Carlton's feedback to Emily, this time to consider how he uses the feedback to model self-regulation for her:

Mr. Carlton:	*Let's look at what you did here . . . OK. Tell me about how you set up your equation.*
Emily:	*Well . . . you have to add up the different numbers to equal $25, and there are songs and an album so you need all the money for the songs and the money for the album.*
Mr. Carlton:	*OK. It looks like you're thinking about adding up some amount for songs and some amount for album, and adding them up to equal $25, so that part of your thinking is correct. You haven't quite translated that into a correct equation yet, though; which of our success criteria is that?*

He both directs her attention to the success criteria as part of his feedback (addressing "What goals am I aiming for?"), and he points out the ways in which she is thinking correctly and is explicit about where her work does not meet the success criteria (addressing "Where am I currently in relation to those goals?").

Emily:	*(Looks up at the board) Um . . . the first one?*
Mr. Carlton:	*Right. Maybe it would help to think about the cost of just 2 or 3 songs and the expression you would use to figure that out. How could you use that thinking to revise your equation? Think about that, and I'll check back with you in a few minutes.*

He also gives her a suggestion to head her in the right direction without completing the problem for her. (This addresses the question, "If I have not met the goals, what do I need to do next to be able to meet them?")

Notice that the example of Mr. Carlton's feedback illustrates giving formative feedback to an individual student, and we've shown this to illustrate an example that contains all the elements of effective formative feedback. However, it's unrealistic to think that a teacher could provide individual feedback to *every* student. Formative feedback can sometimes be delivered quite effectively to a class as a whole, and the choice about providing individual, small-group, or whole-class feedback depends on your need for the lesson. We discuss this in greater depth later in the chapter as we talk about the teacher's role in providing formative feedback; for now, keep in mind that when we talk about formative feedback, it does not necessarily imply feedback only provided to individual students.

Characteristics of Formative Feedback

As teachers, we constantly provide feedback to students about their work; written comments on work, results of quizzes and tests, and responses to students' comments or questions in class all provide a wealth of feedback to students. However, some people are surprised to learn that not all feedback is beneficial. Research has shown that under certain circumstances, providing one kind of feedback can be instrumental in moving students' learning forward, yet research also shows that providing a different kind of feedback can actually be an impediment to students' learning (Hattie & Timperley, 2007). This research shows that

> *the least effective type of instructional feedback [for improving student achievement] concerns the student's self as a person. This type of feedback is generally a global statement about the student, ("good girl," "great try," etc.) rather than the student's performance on a task. Feedback at the self level must improve the student's investment of effort or attitude toward learning in order to make an impact. However, this type of feedback usually has little instruction-related content and therefore fails to affect achievement. In particular, praising students has been shown to have little to no effect. (Center on Instruction, 2008, p. 2)*

This doesn't mean that you should no longer give praise, for example, or offer any motivational comments to students—you certainly should work to build positive relationships with your students and praise them for achievements. Praise serves a different purpose from formative feedback, as does pointing out calculation errors, and so these kinds of feedback are not considered *formative feedback*. In this chapter, we focus solely on feedback intended to move students' learning forward toward the learning intention.

Of the various characteristics of formative feedback, we have chosen to focus on three in particular that we consider essential. Feedback that is formative

1. references the extent to which the success criteria are met (including those fully, partially, and not yet met) and therefore remains focused on the learning intention for the lesson;

2. provides guidance about what the student needs to do next to move closer to reaching the learning intention; and

3. is usable by the student.

Mr. Carlton's feedback for Emily illustrates each of these characteristics. Before we look at other characteristics in his feedback, though, let's look at contrasting feedback given by Mrs. Hopkins in a similar situation. Her feedback is missing each of the characteristics.

On e-Music, you can download an entire album for $12, and individual songs are $1.25 each. Jake has an e-Music gift card with $25 on it. If he buys one album, how many additional individual songs can he buy?

1. What should the variable represent in this problem?

2. Write an algebraic equation that could be used to help you solve this problem.

Mrs. Hopkins is looking for the equation $1.25x + 12 = 25$. She walks up to Eric's desk and sees the following work on his paper:

$$x = songs$$

$$12x + 1.25 = 25$$
$$-12 \qquad\qquad -12$$
$$\overline{\qquad\qquad\qquad\qquad}$$
$$x + 1.25 = 13$$
$$-1.25 = -1.25$$
$$\overline{\qquad\qquad\qquad\qquad}$$
$$x \qquad\quad = 11.75$$

$\$11.75$

Mrs. Hopkins:	*Let's look at what you did here. (Looks at his work.) OK, so first of all, this should say x = number of songs, not just x = songs, right? Remember what we talked about yesterday?*
Eric:	*(Erases* songs *and writes in* number of songs.*)*
Mrs. Hopkins:	*OK. (Pauses) Let's think about this. Does it make sense that Jake would have $11.75 after buying a $12 album, then buying some songs?*
Eric:	*I wasn't sure what the equation should be, so is the equation wrong?*
Mrs. Hopkins:	*Yeah, your equation's wrong, so let's fix it. (She takes his pencil.) What would you need to multiply by x? It's not 12, that's the cost of the album.*
Eric:	*So it would have to be $1.25. (Mrs. Hopkins writes 1.25x on his paper.) Then is it plus 12?*
Mrs. Hopkins:	*Right! (She writes "+ 12" after the 1.25x on his paper.) OK, you can do it from here, right? I know you're a smart guy. You just need to finish your equation, then try solving it again, and check your answer to make sure it makes sense. Keep up the good work.*

Now let's review the characteristics with both Mr. Carlton's feedback and Mrs. Hopkins's feedback in mind.

1. **Formative feedback references the extent to which the success criteria are met (including fully, partially, and not yet met) and therefore remains focused on the learning intention for the lesson.**

   Mr. Carlton's feedback to Emily remains focused on the work he sees in front of him, as he references the fact that she is using correct thinking although her equation does not yet represent the relationship in the word problem. He points her back to the success criteria, and then he gives his feedback in relation to one of them. In doing so, he is underscoring the message that Emily's work today is focused on reaching the learning intention.

   Mrs. Hopkins makes no mention of either the success criteria or the learning intention, leaving Eric to assume that the goal for the problem is to get the right answer, rather than to learn about the correspondence between word problems and equations. She also focuses on Eric's work, but the feedback she provides is not really information that is helpful to Eric in taking steps to adjust his current thinking—it may help him correct his work to get the right answer, but it's unclear whether her feedback helped Eric revise his incorrect thinking that led to the mistakes in the first place. Rather, the series of questions she asks lead him to the correct answer, reinforcing the message to Eric (perhaps inadvertently) that the goal for the problem is to get the right answer.

   Looking back at the self-regulation diagram, Figure 4.2, we can see that in order for students to know what to do next to move their learning forward, they need information about how their work and their thinking compares to the success criteria and therefore to the learning intention. An important part of that information includes identifying where their understanding is on track and what they have done correctly (and should therefore continue to do). Thus, for the student, knowing what he or she has done well is at least as important as knowing what he or she still needs to work on. (Notice that this aligns with the recommendation in Chapter 3 to "Elicit evidence about correct thinking and correct responses as often as you do about incorrect thinking and responses.") In addition, many of the teachers in our professional development program saw that when their feedback began with providing confirmation of what students *did* understand, the students were more receptive to the feedback. Notice how Mr. Carlton begins his feedback to Emily by giving her information about how she has (partially) met the first success criterion when he says, "*It looks like you're thinking about . . . so that part of your thinking is correct.*" Mrs. Hopkins's comments, although with good intention to help Eric improve his work, focus exclusively on what he has done wrong and needs to correct, giving him no information about the fact that he too has correctly identified the variable and has chosen the correct operation.

TEACHERS' VOICES

If I notice the work that the student does well, then the student is more apt to be open to rethinking the problem versus shutting down. I know as a student myself, I felt embarrassed when I didn't understand the work. It is so much easier to be engaged and excited to clarify your misconception when approached from the point of what you *do* understand.

---

2. **Formative feedback provides guidance on what the student needs to do next to meet the success criteria.**

This characteristic has two considerations within it: the feedback provides explicit next steps in the form of a hint, a cue, or a model that the students can use, and that hint, cue, or model aims to provide neither too much nor too little information.

- *The feedback describes explicit next steps in the form of a hint, cue, or model that the student can use to move closer to meeting the success criteria.*

Without the opportunity for a student to act on feedback, the feedback has little impact on a student's learning—and to be able to take this action, students need to understand what that appropriate action is. Formative feedback conveys to the student not only the message "Here's what you did right, and here's where you have not yet met the success criteria," but also includes "Here's what you can do to move closer to meeting the success criteria."

An important feature of formative feedback is that it leaves the intellectual work in the hands—and the brains—of the student and helps students understand what to do next to build that understanding for themselves. Using a sports analogy, having someone explain to you how to throw, kick, or hit a ball in a particular way goes only so far in helping you actually learn how to do it well. No one can do that learning for you, and useful feedback in formative assessment leaves room for that learning by being provided in the form of a hint, a cue, or a model. Here are a few examples of hints, cues, or models, some of them more general, some of them more specific to a particular problem:

- Look at the problem we did yesterday. Think about how the approach to that problem could help you with this one.
- Go back to your math dictionary. How did you summarize in your own words what this word meant? Think about the definition and how this information might be helpful here.
- Try finding the cost with easier numbers such as 6 cans for $2. Use this to help you find a different way to compare the ratios.
- Remember that we also have used an area model to think about how to multiply the various terms in both expressions. Think about how you could use an area model to figure out the product of these two expressions.

- *How could you use someone else in the class to help you figure out some possible next steps? Take a look at the list of people on the board who said they felt ready to offer help to other people, and see who you could talk to.*

Mr. Carlton, for example, talks to Emily about using smaller numbers to try setting up an easier equation, and he suggests using that example to help her revise her current equation and meet the first success criterion, "I can set up appropriate equations for word problems." He concludes by suggesting that she think about using simpler numbers to develop an expression and use that to help her refine her work on the problem; the next steps are clear for Emily and are within her ability to carry them out. Most important, it is Emily who is left to take the next steps.

Mrs. Hopkins also provides Eric with lots of feedback about what to do next to revise his work (but not his thinking), but her feedback differs in that she spells out step by step what Eric needs to do. In the end, Mrs. Hopkins has successfully solved the problem, but there is no reason to believe that Eric would necessarily know what incorrect lines of thinking to avoid in subsequent problems.

- *The feedback provides not too much or not too little guidance to the student.*

If feedback attempts to provide too much guidance, there is nothing left for the student to do or learn. Mrs. Hopkins erred on the side of providing too much information and doing the intellectual work for Eric so that there was nothing left for him to do but make the necessary written corrections to the equation. Although she did so with the intent of helping show Eric the line of thinking that would help him solve the problem, it is still her thinking, not his.

If feedback provides too little guidance, then a student is left unsure about appropriate next steps to meet the learning goal. Providing a hint can sometimes come out sounding too vague for a student to find useful. For example, a teacher can sometimes assume that a student will be able to connect the dots in a teacher's explanation or hint, when that assumption is still premature. Consider a teacher who asks several pointed questions such as "So what would you need to subtract from both sides of the equation?" or "How can you rewrite your equation so that you have the total amount of money for the songs plus the total amount of money for the album?" and then looks knowingly at the student and leaves the student to complete the work. If the student is unable to connect the dots between the hints provided by the teacher, the feedback is useless.

Images of Goldilocks and the Three Bears may come to mind as teachers try to figure out what is the just-right amount of feedback to provide to a student. Finding this just-right amount is a continual judgment call on the part of the teacher and is what makes providing effective formative feedback trickier than it may seem at first. Mr. Carlton attempts to strike the right balance when he suggests to Emily, *"Maybe it would help to think about the cost of just 2 or 3 songs and the expression you would use to figure that out. How could you use that thinking to revise your equation?"*

3. **Formative feedback is usable by the student.**

This characteristic also can be broken down into multiple considerations around how you give feedback in a way that students actually can use it to move their learning forward.

- *The feedback is worded in a way that students can understand it.*

   In order for students to use feedback they receive, they need to understand it, both in terms of the vocabulary you use and of the meaning of the feedback as a whole. However, keeping feedback understandable does not mean that academic language is to be avoided. Sometimes, the purpose of a lesson can include gaining familiarity with mathematical academic language, such as the meaning of ratio or function, and learning to be fluent and facile with the language and symbols of mathematics is an important goal for all students. Rather, it means that the student simply needs to be able to make sense of the feedback he or she is receiving in order to act on it effectively.

- *The feedback is concrete and specific.*

   In order for students to use feedback they receive, it also needs to be concrete and specific. This may seem like a contradiction to the idea of providing a hint, cue, or model, but let's look at Mr. Carlton's suggestion: *"Think about the cost of just 2 or 3 songs and the expression you would use to figure that out. How could you use that thinking to revise your equation?"* This provides an example of feedback that is concrete enough to act on, without spelling out the specific next steps to follow, leaving some thinking for Emily to do. An example of unspecific or nonconcrete feedback would be feedback typical of the kind some of us have received on English papers, "Say more here!" or on math tests, "Check your numbers!" Dylan Wiliam, a British mathematician and education researcher, cleverly captured the need for concrete and specific feedback when he said

   Feedback is formative only if the information fed back to the learner is used by the learner in improving performance. If the information fed back to the learner is intended to be helpful, but cannot be used by the learner in improving her own performance it is not formative. It is rather like telling an unsuccessful comedian to "be funnier." (1999)

- *The feedback is prioritized.*

   In order for students to use feedback they receive, it needs to come in dosages that are manageable by students. Carefully giving students a long list of all the things that need correction in a particular math problem can serve to overload many students and leave them overwhelmed and checked out. Instead, formative feedback helps students move closer to reaching the learning intention by taking steps at a time, rather than one single giant leap. The number of steps—and the size of the steps—certainly depends on the individual student.

An analogy can be drawn to the use of focused correction areas that some teachers use to teach writing, in which the teacher identifies several points for feedback in a piece of student writing and provides feedback only on those points. One purpose in doing so is to avoid overwhelming the student with multiple corrections needed for spelling, for punctuation, for sentence structure, for paragraph structure, for vocabulary usage, for voice, and for developing ideas within a paragraph and across the entire piece. While all those writing elements may need further work, the teacher focuses the feedback on a selection of writing elements in order to help the student focus his or her revision efforts to those areas most pertinent to the goal for the lesson.

Limiting feedback to the information that references the mathematical success criteria—and choosing which of those success criteria to comment on at a time—is the mathematical equivalent to this process. If there are two or three success criteria, the feedback may only address one of those at a time. Notice that Mr. Carlton chooses only to focus his feedback to Emily on the equation, even though there are other errors (she identifies songs, rather than number of songs, as the variable; she also solves the equation incorrectly). Mrs. Hopkins, on the other hand, works step by step with Eric to make all the necessary corrections, which is useful to him only if he can effectively keep track of them all and keep them in mind for subsequent problems. Prioritizing feedback does not mean that the other mistakes or issues in a student's work remain entirely ignored; it simply means that they may be best addressed at a different time, in order to help the student focus on the key learning stated in the learning intention for that day. This approach can be particularly effective for struggling students who often have difficulty pinpointing a specific area of difficulty for themselves and may tend to generalize their difficulties to a broad statement of "I can't do math."

- *The feedback is timely.*

Formative feedback that a student can use to modify his or her learning is only effective if the student receives it while the learning is still underway. For example, a teacher who collects exit tickets from students after class on Monday is ready to start class on Tuesday with some feedback for the whole class, based on the responses to their exit tickets; this feels timely for students because the learning intention, the exit ticket, and maybe their responses as well are still fresh in their minds and relevant to their current learning intention. For this reason, we do not consider grades and comments on a test or quiz that is passed back at the end of a chapter or a unit to be formative feedback, although it certainly is feedback on the level of mastery of the material.

Grant Wiggins makes a distinction between "timely" and "immediate" feedback in a humorous way when he says,

In most cases, the sooner I get feedback, the better. I don't want to wait for hours or days to find out whether my students were attentive and whether

they learned, or which part of my written story works and which part doesn't. I say "in most cases" to allow for situations like playing a piano piece in a recital. I don't want my teacher or the audience barking out feedback as I perform. That's why it is more precise to say that good feedback is "timely" rather than "immediate." (2012)

The following summarizes the key characteristics of what constitutes formative feedback. As we continue to define what formative feedback is, and how it is used, we next look at two contexts for giving formative feedback: directed toward the whole class during instruction or given to an individual student orally during instruction or in writing on an assignment.

Three Key Characteristics of Formative Feedback

1. Formative feedback references the extent to which the success criteria are met (including fully, partially, and not yet met) and therefore remains focused on the learning intention for the lesson.

2. Formative feedback provides guidance on what the student needs to do next to meet the success criteria.

   a. The feedback describes explicit next steps in the form of a hint, a cue, or a model that the student can take to move closer to meeting the success criteria.
   b. The feedback provides not too much or not too little guidance to the student.

3. Formative feedback is usable by the student. In particular, the feedback is

   a. worded in a way that students can understand it,
   b. concrete and specific,
   c. prioritized, and
   d. timely.

Reference Resource: Go to Resources.Corwin.com/ CreightonMathFormativeAssessment to access this resource.

 Chapter 4 Summary Cards. This printable file provides Summary Cards that include characteristics of formative feedback.

Learning Resource: Go to Resources.Corwin.com/ CreightonMathFormativeAssessment to access this resource.

 Feedback Characteristics—Content. This interactive web page provides practice deciding whether feedback examples meet the characteristics related to content.

Feedback Characteristics—Usability. This interactive web page provides practice deciding whether feedback examples meet the characteristics related to student usability.

Whole-Group Feedback Versus Individual Feedback

Earlier in this chapter, we talked briefly about the use of whole-class feedback, not just individual student feedback. To better understand what formative feedback might look like for the whole class, let's consider an example.

Example: Mrs. Rice liked to use questions for her learning intentions. For one lesson, the learning intention and success criteria in Mrs. Rice's math class were the following:

> LI: How does a scatter plot provide information?
> - SC1: I can describe the steps to create a scatter plot.
> - SC 2: I can create a scatter plot for a given set of data.
> - SC 3: I can explain how parts of the scatter plot relate back to the situation it represents.

At the end of the lesson, she collected student papers in which they had created two different scatter plots, one with support from her and one on their own.

The next day, before handing back the papers, she said, "When I looked over your work, I saw that many of you were able to meet at least one of our success criteria, but there were none that everybody met. So that tells me that we still need to do some more work understanding how to construct a scatter plot, but that each of you needs to focus your work on different success criteria.

"I noticed several common issues that need fixing, and I've listed those issues on the table here." Mrs. Rice projected the following table for the class to see. In the table, she described difficulties from the students' work written next to the second success criterion with some examples of the associated difficulties.

I can create a scatter plot for a given set of data.	*Scale Difficulties* - *Not starting at 0* - *Too large a range*
	Graph Difficulties - *Labeling intervals in spaces rather than lines* - *No labels/missing labels* - *Confusing coordinates (which is horizontal, which is vertical)*

(Continued)

(Continued)

I can explain how parts of the scatter plot relate back to the situation it represents.	*Explanation Difficulties* • *Not including all the parts (categories and scale)*

She repeated this for the third success criterion. (The first success criterion—describing the steps to create a scatter plot—was not addressed here, because the task resulting in this work did not include a question to address it.)

She went on. "When you get your paper back, use the list to determine which mistake or mistakes you may have made in order to revise your work so that it meets the success criteria."

After clarifying each of the issues, she returned their papers and directed students to find what they did correctly and then to find and fix their mistakes. Mrs. Rice walked around supporting students as they worked and asking those who finished to help their group members who had questions. After collecting the papers, she introduced the new learning intention and success criteria for the rest of the day's lesson.

There are several key points about providing whole-group feedback that are worth noting here:

- Formative feedback references the success criteria. For a whole class, this can be done in a public way as Mrs. Rice has done here, showing which parts of the feedback relate to each success criterion.
- Formative feedback involves making clear how the work does and does not meet the success criteria, but with a whole class full of students, it's not possible to do this for individual student's work in this way. With whole-class formative feedback, the purpose then becomes to reflect back to the class where they are generally as a whole and what the class needs to work on next. Even though students are not getting individually tailored feedback, the benefit to students with whole-class formative feedback is that it helps make public the direction the teacher is taking the learning, *and why*. In effect, whole-class feedback serves the purpose of clarifying the learning intention and the focus of the lesson for students by providing a rationale.
- Formative feedback provides clear and enactable next steps for students, but to do this at a whole-class level usually means giving students some kind of self-assessment exercise to do in which they have to determine which pieces of the feedback apply to them. Mrs. Rice does this by asking students to review the list of issues, to figure out which issues apply to them, and to make revisions accordingly. If she felt her students needed it, she could have gone a step further toward helping them learn to self-assess by modeling the process for them with an example. For instance, she could have used a fictitious piece of student work and modeled how to evaluate the work against the list of issues using a *think-aloud*, talking aloud about how she feels the work compares to the list of issues. With

individual formative feedback, the person providing the feedback can be specific about how the work does and does not meet the success criteria, so this kind of self-assessment exercise is not always necessary (though it could certainly be included).

Another Example of Whole-Group Formative Feedback

Let's consider another brief example of whole-class feedback. As you read through the example, look for the ways in which the teacher

- reflects back to the class as a whole where they are in their learning, with specific reference to the success criteria;
- creates an opportunity for students to determine what parts of the feedback are pertinent to their individual work; and
- creates an opportunity for students to use the feedback to revise their work.

Mr. Patari works down the hall from Mrs. Keller (whom we met in Chapter 3), and like her, he is focusing his eighth graders on rate of change. His learning intention and success criteria for this lesson are:

LEARNING INTENTION AND SUCCESS CRITERIA

Learning Intention: I will understand how "rate of change" appears in different representations.

Success Criteria:

1. I can identify the rate of change in a problem.

2. I can explain or show where to look for the rate of change in tables of values.

3. I can explain or show how the rate of change appears on a graph.

Students are working on some problems about identifying rate of change in both tables and graphs; one problem from their handout is shown below.

Colin likes to go ice skating. At the skating rink, he can rent skates for $5, then pays $3 per hour for each hour he skates.

1) Complete the table of values:

Number of Hours	Cost for Renting Skates
0	$_____
1	$_____
2	$_____
3	$_____
4	$_____
5	$_____

(Continued)

(Continued)

2) What is the "rate of change" in this problem? (Remember to use our sentence frame below):

For every _____ that you increase/decrease, the _____ increases/decreases by _____.

3) Find the graph on the back of this page that matches your rate of change description and write the letter of the graph here: _____.

4) Explain how your graph matches your rate of change of description.

As Mr. Patari walks around his classroom talking to students, he takes time to record on the board some of the comments he has been providing to different students he talks to. His list consists of the following:

1. When you describe the rate of change, thinking about dependent and independent variables might help.

2. Rate of change compares two different quantities.

3. See if you can recall our discussion earlier this week about how a rate is a ratio. How can you make ratios from your table of values?

4. We've talked about where dependent and independent variables appear on a graph. How could that help you find where (or how) the rate of change appears on a graph?

Each time he adds a new comment to his list on the board, he notices that many of his students look up, read the comment, and look back at their own papers. After students have had some time to work on their own, he is ready to give the class some collective feedback and to encourage them to do some self-assessment using the success criteria.

Mr. Patari: *OK, let's come back together to talk about what you've been working on. Take a look at our success criteria again: there are three things we're looking for today. Think for a moment about which of these you feel most solid on and which one or ones you feel less sure of. I've got four comments on the board here that are comments I've been making to some of you as I've been talking to you; these comments may provide some idea of what to think about for the problems you're unsure of.*

I'm really pleased to see that everyone I spoke to could tell me something about what a rate is, so many of you are meeting our first success criterion. (Points to it.) However, if you're still not sure you're meeting it, then keep in mind these first two comments on my list; we've been talking about dependent and independent variables over the past couple days so think about that relationship to help you think about rate of change. And remember that a rate involves describing two quantities.

I also noticed that many of you seem most comfortable working from a table and identifying a rate from a table of values—but not all of you. (Points to the second success criterion.) So look at your own work, and decide whether you feel like you are meeting this one. You may want to keep in mind Comment 3; think about what you know about the connection between a ratio and a rate, and think about how ratios appear in your table of values. Comment 2 might help some of you as well there.

(Points to the third success criterion.) The remaining evidence we're looking for is how you can explain or show how the rate of change appears on a graph. Again, think about whether you feel you're meeting this one. If not, Comment 4, and maybe also Comment 1, may help you there. Think about our work with dependent and independent variables; how might they appear in rates, and how might they on a graph?

Mathematics Background:
Dependent and Independent Variables

It's common in mathematics to describe a relationship between two sets of values in which one of the sets of values depends on what happens to the other set of values. For example, for some jobs, the amount of money you earn depends on the number of hours you've worked; or, when the number of cells in a biology experiment is doubling every so often, you can predict the number of cells based on the amount of time that has passed. One of the sets of values is dependent on the other and is represented by the dependent variable. The remaining set of values is represented by the independent variable. On a graph, the independent variable is commonly put on the x-axis, and the dependent variable on the y-axis. (In both of the examples, the independent variable is time; the dependent variable is amount of money in one example and the number of cells in the other.)

The rate of change for a relationship is described by how much the dependent variable changes according to how much the independent variable changes, for some interval of values in the independent variable. For a linear relationship, including proportional relationships, this rate of change is constant for any interval. For example, someone making $20 per hour makes that same rate whether they work 1 hour or 8 and whether you look at the first hour or the seventh.

In this example, Mr. Patari reminds students of the success criteria and points out how different parts of his feedback relate to the criteria. He has been working on teaching his students how to self-assess their work in relation to the success criteria, and he uses this as another opportunity to give them practice doing so. Finally, they have the opportunity to act on the feedback as they keep working. Because he is choosing to provide some feedback while they are in the midst of working on the problems, his students have information they need to make their own midcourse corrections as they are able. Mr. Patari can now continue to circulate and keep an eye out for those students who are able to make use of the feedback to improve their work and those students that need some additional support from him.

Remember, there are many kinds of feedback that students receive, but a key distinction about formative feedback to support the formative assessment process is that students use the feedback to adjust and revise *their own learning*. Thus, not only does formative feedback help students correct their thinking, but it also helps them learn to evaluate their own work in relation to the learning goal and success criteria and take steps to move their learning forward.

Creating clear, usable formative feedback for a student takes practice. A major benefit of making that effort is that you're buying yourself an opportunity to build students' capacity to self-regulate their own learning.

■ USING FORMATIVE FEEDBACK IN YOUR CLASSROOM

Like many other elements of formative assessment, the idea of incorporating the critical aspect of formative feedback into your classroom may seem pretty straightforward—as teachers, we're accustomed to giving feedback to students—but it takes some time and practice to learn to construct feedback that fits the characteristics, to judge when best to deliver it, and to build in opportunities for students to respond to it. In this section, we discuss your role in formative feedback as well as your students' role and what you can do to support your students learning to step into their role.

Feedback or Instruction?

In the previous chapter, we discussed four different responsive actions that a teacher can take, depending on his or her interpretation of the evidence about students' reasoning and understanding, one of which was providing formative feedback:

- Gather more evidence.
- Provide formative feedback.
- Provide further instruction.
- Move on.

As we said in Chapter 3, while we list these as four separate responsive actions, in the reality of the classroom, the lines between these responsive actions is often blurred, and a teacher will often include more than one in an interaction with a student. Providing formative feedback is particularly easy to confound with providing further instruction, since there are certain definitions of feedback for which feedback and instruction would be one in the same. For instance, in our earlier examples with Mr. Carlton and Mrs. Hopkins helping their students develop equations for word problems, Mrs. Hopkins was trying to give feedback that was in the form of further instruction as she walked the student through the correct steps to solve the problem.

However, formative feedback is *not* the same as providing additional instruction to a student who is still unclear about a mathematical idea, because within formative assessment, these two responsive actions serve slightly different purposes. The difference lies in how big the gap is between a student's current understanding and understanding of the learning intention. If a student has elements of correct thinking that a teacher can build on, then a teacher can provide effective formative feedback by building on that correct thinking and help the student close the gap. If the gap is too large, then further instruction is more appropriate.

Consider a student, Anna, who initially encountered ratio tables such as the following as a tool for learning about equal ratios.

Money Earned	$8	$40	?	$88	$200	$400
Hours Worked	1	5	10	?	25	?

After working for some time with these ratio tables, she encounters input/output tables in relation to line graphs. Although she has seen many examples of tables that are not ratio tables, she begins to assume that any table relating two variables, such as each of the tables shown below, must have values that are proportional. While the first two tables happen to have proportional values, the third does not. (See the mathematics background on proportional values in Chapter 3.)

Input	Output
4	10
6	15
8	20
10	25
12	30

x	y
0	0
3	6
6	12
9	18
12	24

Time (hr)	Distance (mi)
1	100
2	150
3	200
4	250
5	300

One day in class, Anna's teacher lays out the following learning intention and success criteria:

LEARNING INTENTION AND SUCCESS CRITERIA

Today we will learn how different representations show proportional relationships.

How you will know if you've met the learning goal:
- You can show a proportional relationship using at least two different representations.
- You will explain how each representation you created shows a proportional relationship.

Anna makes the Time/Distance table shown above. When her teacher stops at her desk to ask her about her table, Anna says that it's proportional because the numbers are going up by the same amount in each column. Her teacher recognizes that Anna has elements of correct thinking—that proportional relationships can be expressed as tables with values that increase or decrease at a constant rate—but that she is misapplying that knowledge in this situation. Anna probably doesn't need further instruction, because her thinking is good. Her teacher decides to give formative feedback instead:

"You're correct that proportional relationships can be shown in tables with an input and output, and the numbers go up by the same amount in each column." (Here the teacher is pointing out her correct thinking in relation to the success criteria.)

"But remember that we talked about how proportional relationships have a constant factor–a number you can multiply each input number by to get the output number–and that factor doesn't change. I don't see that constant factor in the table you've created." (Here, the teacher is pinpointing ways in which she has not met the success criterion.)

"See if you can revise your table using this idea of a constant factor." (Finally, the teacher is providing a hint to help Anna meet the success criterion.)

With this formative feedback, Anna has specific information about where her thinking is correct, where it is flawed, and a concrete next step to take to try to meet the success criterion.

The Teacher's Role Before the Lesson: Planning for Formative Feedback

Though much of the use of formative feedback in formative assessment takes place in the moment or is generated for students' work after a lesson, there are some ways in which planning for formative feedback can strengthen your lesson. That's why we focus on doing so as our first recommendation in this chapter:

> **Recommendation 1: Include planning for formative feedback as part of your lesson planning.**

At first glance, it may seem impossible to plan for formative feedback: How can you know what you will need to say, and to whom, about what? Although you can't necessarily plan *what* you will say in your feedback to different students, you *can* plan the focus of your feedback, the way in which you or students' peers might give it, and the way in which students might act on it. Let's look at each one of these elements of your planning.

1. **Plan the focus of your feedback by drafting a possible model response.**

Planning the focus of your feedback consists of clarifying for yourself *as precisely as you can* what the evidence of meeting the success criteria will look like or sound like. The clearer you are about this for yourself, the more easily you will be able to provide specific feedback to students that helps them move their learning forward. In a sense, you are drafting a possible model response for meeting the success criteria. Articulating a model response for yourself has two big advantages:

- *It causes you to clarify for yourself precisely what the goal is for a student's response, so that you can more easily evaluate a student's actual response.*

  The benefit to you is that you can be much clearer about what you're listening and looking for in student responses, and therefore you can be much clearer in providing feedback to a student about how their response does and does not yet meet a success criterion.

- *It gives you an opportunity to check how well you've written your success criteria.*

  If your success criteria are written well, a model response will go far beyond simply providing a correct answer. Some of the teachers in our professional development program found that when they started to draft a possible model response of what they hoped to hear, their success criteria needed to be clearer. As your learning intention, success criteria, evidence, and formative feedback all start to correspond more clearly to each other, student learning will gain momentum.

But what about all the other possible responses you might get from students that would be equally acceptable? And what about the fact that it's often unlikely that you'll hear this model response from many of your students? It's important to note the primary reason you create a model response: in doing so, you will clarify *for yourself* what it is that you will be listening and looking for, for each of your success criteria. For that purpose, you don't need many possible responses, nor do you even need to feel like you've chosen a best one. It doesn't matter that students may not be able to produce this model response, at least not right away. You simply need to come up with one possible model response for the goal of creating this clarification. Let's look at an example of planning a model response to illustrate this idea, using Mr. Patari's class again.

Based on the evidence he gathered in today's class, Mr. Patari decides to stick with the same learning intention and success criteria for tomorrow, but to put more of his focus on helping students meet the third success criterion:

LEARNING INTENTION AND SUCCESS CRITERIA

Learning Intention: I will understand how "rate of change" appears in different representations.

Success Criteria:

1. I can identify the rate of change in a problem.

2. I can explain where to look for the rate of change in tables of values.

3. I can explain where to look for the rate of change in graphs.

Focusing on that third success criterion, he asks himself a question:

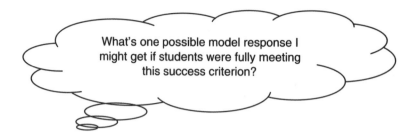

What's one possible model response I might get if students were fully meeting this success criterion?

After some thought, he comes up with this possibility:

In a line graph, I would look for the rate of change by looking at any two points and figuring out how much one variable increased or decreased in relation to how much the other variable increased or decreased between those two points. The rate would then be a ratio of those changes. On the graph, this is the same as looking at the slope of the line.

Mr. Patari knows full well that it's highly unlikely any of his students will produce this model response. His students don't need to say these exact words or make every one of his points. Recall that the purpose of this ideal response is not to articulate what he thinks his students are likely to say but instead to articulate for himself what it means to fully meet the success criteria. Having gone through this mental exercise, Mr. Patari can now identify the points that he feels are most vital for students to articulate in some way:

- Both quantities in a rate are important.
- You can find the rate from any pair of points on the line.
- The change, or difference, in each quantity is important to the rate of change.
- The rate of change is a ratio of those differences.

He also realizes that he would like to hear students connect the rate of change to the slope (see the mathematics background on this content in Chapter 3), but this is more than his current learning intention requires. He might think of this as a stretch goal for students who are getting the other points and who are ready for the next step.

Having this clarity not only supports his use of formative feedback but also his communication of the learning intention and success criteria to students and his elicitation and interpretation of evidence as well.

A Note on a Common Pitfall: "Cutting Corners" on Creating a Model Response

Creating a model response is remarkably tricky to do, though it can seem like it would be quite easy. When the teachers we worked with were first learning to do this, it was not uncommon for them to create a model response that was much less specific than it needed to be, such as

- "Students will tell me something about slope and something about rate";

- "They'll talk about how they see the relationship between a rate and a ratio"; or even just
- "Slope, rate, variables."

Be wary of talking *about* what the students will say, rather than actually creating the possible model response, or just listing topic headings. Both of these are tempting ways to cut corners on developing a possible model response, but they gloss over the point: What is that "something" about slope and rate? What relationship is it that you want them to see? What about slope, rate, and variables should they say? Taking the time to draft the response will pay off in big ways in your formative assessment implementation.

 Planning Resource: Go to Resources.Corwin.com/ CreightonMathFormativeAssessment to access this resource.

 Formative Assessment Planner Templates. This printable file provides two versions of templates to help you create a formative assessment plan for a lesson. One version includes a portion in which you can record model responses for the success criteria.

2. **Plan the way in which you or students' peers might give feedback.**

Planning the way in which students will receive feedback has two parts to it: How might you provide feedback, and how might you structure the lesson so that students can provide feedback to their peers. Think about what opportunity could be created during the lesson for students to usefully receive some feedback. These opportunities can occur at various times during a lesson. Sometimes it may be appropriate at the start of a lesson in response to something you noticed during discussion from the day before or from students' exit tickets from the end of the previous lesson; other times, it may be appropriate in the middle of a lesson to receive some feedback on their work in progress. This is often given informally as you observe their work and see moments to help redirect their efforts or see that many of the students are well enough along that they can provide informal help to their peers. More structured moments are helpful, especially when students are giving peer feedback; see the Resources section for some example templates. We say more about helping students learn to give peer feedback in Chapter 5, "Developing Student Ownership and Involvement in Your Students."

 Classroom Material: Go to Resources.Corwin.com/ CreightonMathFormativeAssessment to access this resource.

 Feedback Templates. This printable file includes templates that you can use to provide feedback to your students, and that they can use to provide peer feedback.

3. Plan how students might act on feedback.

Planning how students might act on feedback involves making sure that there are opportunities in the lesson for students to do something with the feedback they receive. If students don't have time and opportunity to respond to feedback that they've received, many are likely to ignore it. They may even feel like it must not have been important in the first place, since they have no accountability for looking at it or responding to it. With the demands of addressing lots of content during class time, it's easy to shortchange this important last step. However, without any opportunity to act on the feedback they receive, students may never adjust their thinking or skills, and your feedback will be ineffective.

The following questions can help you think about how you want to plan for students to respond:

- When will they respond? During the lesson, or after the lesson as homework?
- Will they work individually or in small groups?
- What kind of structure do you want for the response? For example, will they revise their work using the feedback as a guide? Will they try a similar problem so they can apply the feedback to a new situation? Or will they write a reflection explaining how the feedback can help them in the future?

Providing templates and structures for responding to feedback can be helpful, especially when students are relatively new to the expectation that they must respond to it.

 Classroom Resource: Go to Resources.Corwin.com/ CreightonMathFormativeAssessment to access this resource.

 Feedback Response Strategy. This printable file describes a strategy that you can give your students to help them respond to feedback.

The Teacher's Role During the Lesson

Much of a teacher's role in providing formative feedback to students takes place during the lesson, in response to students' evolving understanding of the learning intention. As we consider the various recommendations for formative feedback during a lesson, we'll illustrate them using example responses from three different teachers.

A sixth-grade class is working on ratios and has the following learning intention and success criteria:

LEARNING INTENTION AND SUCCESS CRITERIA

Learning Intention: Ratios can be used in more than one way to compare quantities.

Success Criteria:

1. Use appropriate ratios for a given situation.

2. Use multiplication or division to compare ratios.

3. Explain why you can use ratios to compare quantities in more than one way.

In a math task, students had to determine which store had a better deal on cat food: Sam's, which sold 12 cans for $15, or Pam's, which sold 20 cans for $23.[1] One student's work is shown in Figure 4.3, but we suggest you take a moment to solve the problem yourself before proceeding. Note that this task is only part of the lesson and only is intended to address the first two success criteria.

Figure 4.3   Student Work

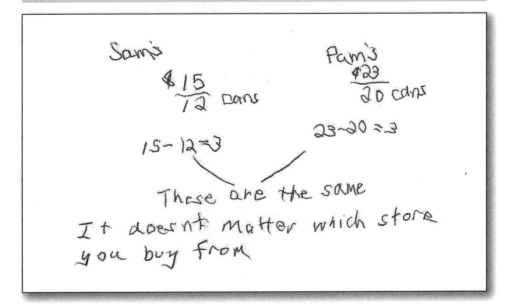

Here are examples of feedback on the student work from three different teachers. Teacher A meets none of the recommendations we provide, while Teacher B meets some of them, and Teacher C meets all of them.

---

1. This task is adapted from *Best Buys, Ratios and Rates* by Bill Jacob and Catherine Twomey Fosnot (2008).

Asia, you're usually my super math student—let's see what you've done. OK, does it make sense that they would be the same? Let's *think* about this: If you have 12 cans for $15, what do 4 cans cost? And if you have 20 cans for $23, what do 4 cans cost?

Teacher A: Meets None of the Recommendations

Your work is very neat and easy to follow, nice job! Our second success criterion said to use multiplication or division to compare the ratios, so I don't understand why you chose to use subtraction. Fractions are another way of showing division, right? So you want to divide 15 by 12, then divide 23 by 20, and then compare those results. Whichever one is smaller will be the better price, right?

Teacher B: Meets Some of the Recommendations

It looks like you used appropriate ratios for the problem, and so you've met the first success criterion. Then you used subtraction to compare the ratios. Unfortunately, subtraction is not the correct operation. Look at our second success criterion that mentions using multiplication or division. Try using multiplication or division to find the cost of one can using easier numbers. See if that helps you find a more accurate way to compare these ratios and figure out the right operation.

Teacher C: Meets All of the Recommendations

> **Recommendation 2: Provide feedback to students that points out what criteria have been met (if any), what criteria have not been met (if any), and a suggestion for a next step to move learning forward.**

The middle part of this recommendation—pointing out what criteria have not been met, if any—is a common perspective for teachers. As teachers, we all want our students to do better, so we tend to focus on what they are missing, with the intention of helping them improve. Since the other parts of this recommendation tend to be overlooked, we want to focus your attention on them in the following paragraphs.

- **Point out what criteria have been met, not just those that are unmet**

Getting feedback on those things you've done correctly is equally valuable to getting feedback on those places that need correction. For many students, it's easier to take in feedback when the first thing you hear is what you did right. This does more than build self-esteem, though; it deepens a student's learning because it articulates for the student where his or her thinking is correct and therefore should be continued. You would expect a sports coach or music teacher to reinforce your good habits for kicking a ball, swinging a club, or holding your instrument correctly; it helps you understand what techniques and habits to continue, not only what to abandon. Why not a math teacher pointing out correct mathematical thinking or procedures? Teacher C models this by starting with a comment on how the student met the first success criterion ("It looks like you used appropriate ratios for the problem, and so you've met the first success criterion"). This is distinctly different from Teacher B complimenting the student about something unrelated to the learning ("Your work is very neat and easy to follow, nice job!").

TEACHERS' VOICES

My kids feel less stress about math. They are focused on the success, thinking, "I can do this one. I've got this." Before I would be focused on a deficit model—what they don't get. As a teacher, I focus on what I have to fix. Now together, we focus some on the positive and the successes in their learning. Instead of a blanket feeling of not being able to do math, they have the opportunity to delineate the parts of their learning and feel success. What they have been successful with can be a jumping off point.

- Provide a suggestion for a next step to move learning forward.

  Remember that formative feedback describes actual next steps that the student can take, is concrete and specific, and is in the form of a hint, cue, or model for students to use to move their work forward themselves. That is, your feedback should include clear, specific steps that, if the student can follow them, will help him or her move closer to reaching the learning intention (not just to the correct answer for a problem).

  Teachers B and C both include explicit suggestions to the student about next steps to take ("So you want to divide 15 by 12, then divide 23 by 20, and then compare those results," and "Try using multiplication or division to find the cost of one can using easier numbers. See if that helps you find a more accurate way to compare these ratios and figure out the right operation"). Both are concrete and specific. They contrast with Teacher A, who is implying next steps but not specifying them in a way that most students could easily act on ("If you have 12 cans for $15, what do 4 cans cost? And if you have 20 cans for $23, what do 4 cans cost?"). After the student answers Teacher A's questions, only some students will know what that means for how to compare the ratios correctly.

  As you try enacting this recommendation, remember the characteristic that feedback provides not too much or not too little guidance to the student, and be careful to keep the intellectual work in the hands of the students. This is where much of the art of effective formative feedback lies, and at the same time, it is probably the most challenging part of this recommendation for teachers to meet.

For example, let's look at Teachers B and C again. Teacher B spells out exactly what to do (". . . divide 15 by 12, then divide 23 by 20, and then compare those results") and leaves only the number crunching for the student to complete; rather than providing feedback as a hint, cue, or model, the feedback comes out as a set of specific directions. Teacher B's effort to be helpful and specific results in little or no learning for the student. Teacher C, on the other hand, provides a concrete, enactable suggestion, but leaves the follow-up intellectual work in the hands of the student. Part of this comes from Teacher C providing a hint or cue ("Try . . . See if that helps . . .").

Learning Resource: Go to Resources.Corwin.com/ CreightonMathFormativeAssessment to access this resource.

Feedback Characteristics—Content. This interactive web page provides practice deciding whether feedback examples meet the characteristics related to content.

Recommendation 3: Keep the feedback specific and focused on the success criteria.

Formative feedback needs to come in dosages that are manageable by students, and being specific in your feedback is one way to keep it manageable. If a student's thinking is at Point A and you want to get the student to Point E by the end of class, it's sometimes easy to fall into the trap of giving feedback on Points A, B, C, and D all at the same time in order to jump right to Point E. Few students will be able to navigate so many steps to get through, and some may feel there's no point in trying if they're so far away from that final point. So consider that you may not need to give feedback on an entire assignment; perhaps it's sufficient to focus your feedback on one success criterion reflected in students' work on one or two key questions that are part of the assignment or on the whole-class discussion that results. This focus on the success criteria helps you keep the feedback specific.

Explicit references to the success criteria also provide a clear connection between the feedback and the success criteria. This connection models for students that they should be looking at the success criteria as they evaluate their own work; it also helps students maintain a focus on reaching the learning intention. Teacher A's comments, with no reference to the success criteria, could leave a student focused on nothing more than the perceived goal of getting the right answer. In contrast, Teacher B references one of the success criteria to give some context for her question about the student's use of subtraction ("Our second success criterion said to use multiplication or division to compare the ratios, so I don't understand why you chose to use subtraction"), and Teacher C references two of the success criteria ("It looks like you used appropriate ratios for the problem, and so you've met the first success criterion. . . . Look at our second success criterion that mentions using multiplication or division").

Another important purpose for focusing the feedback on the success criteria is that this helps you give feedback that is descriptive *of the student's work* and doing so *in relation to the success criteria*. Teacher A mentions the student's answer by saying, "Does it make sense that they would be the same?" but this only implies that the answer is wrong; he or she makes no mention of the work the student did to find the answer and no mention of the success criteria. The teacher doesn't seem to be looking at the student's work at all, only at the answer. Notice that Teacher B is descriptive when he or she says, "Your work is very neat and easy to follow, nice job!" However, being neat is not related to any evidence of learning described by the success criteria, so while the comment may contribute to the student's self-esteem, it will not have any impact on the student's learning of the learning intention. Teacher C's comments remain focused on the student's work and do so in relation to the success criteria.

Again, remember the characteristic that feedback provides not too much or not too little guidance to the student. As we mentioned in the characteristics section, finding that Goldilocks just-right amount is a continual judgment call for the teacher. All judgments about what is appropriate feedback are based on your knowledge of your particular students, and those judgments probably improve as the school year goes on, and you get to know your students better and better. Teacher C seems to provide a good example of feedback that meets this recommendation—specific while still leaving the intellectual work to the student—but whether that's enough (and not too little) for this student depends on the student's learning needs. Similarly, at face value, Teacher A might seem to be providing not enough; depending on what Teacher A knows about the student, though, it might in fact be just right to get the student going in the right direction. The point is to provide enough to keep students moving forward productively in their work independently, though we must recognize that different students may require a different amount of information and guidance.

TEACHERS' VOICES

I find that I often meet with students and provide feedback, but after looking at the recommendations . . . it is abundantly clear that I need to be more precise and concise with what I say. . . . I also noticed that I will meet with students and discuss problems in their work with them, but I often hold their hand too much and don't give them the tools and tips to succeed but rather walk them through it more. I have been attempting to change these practices, and I have noticed significantly improved results in my students' quality of work and understanding. Sometimes it's better to say less and be very concise and precise with your words to allow students the necessary structure to succeed in solving these problems themselves.

Learning Resource: Go to Resources.Corwin.com/ CreightonMathFormativeAssessment to access this resource.

Feedback Characteristics—Usability. This interactive web page provides practice deciding whether feedback examples meet the characteristics related to student usability.

Classroom Resources: Go to Resources.Corwin.com/ CreightonMathFormativeAssessment to access these resources.

Framework for Using LI and SC With Students. This printable file describes a way to help students understand and use learning intentions (LI) and success criteria (SC).

Revisiting Strategies. This printable file describes strategies for revisiting learning intentions and success criteria in the midst of a lesson, including a strategy for providing feedback.

Teacher Summary Cards for the Strategies. This printable file provides large-sized cards that provide step-by-step summaries of teacher moves for the Revisiting Strategies, including a strategy for providing feedback.

Feedback Techniques. This printable file lists strategies for providing feedback that can be found in the book *Mathematics Formative Assessment: 75 Practical Strategies for Linking Assessment, Instruction, and Learning* by Keeley and Tobey (2011). A brief description of each strategy is included.

Classroom Materials: Go to Resources.Corwin.com/ CreightonMathFormativeAssessment to access these resources.

Student Summary Cards for the Strategies. This printable file provides large-sized cards with step-by-step summaries of student moves for the Revisiting Strategies, including a strategy for providing feedback. This can be distributed to each student or pairs of students for use when learning about a strategy.

Feedback Templates. This printable file provides templates that can be used for the feedback techniques described earlier.

> **Recommendation 4: Make sure students understand the feedback they're receiving.**

Simply put, if a student is going to be able to act on feedback, he or she needs to understand it. Perhaps every teacher has encountered the student who, upon receiving feedback, will not want to say that he or she still doesn't understand, so the student just nods at you and then wonders what to do. Although frustrating for teachers, it is not unexpected:

> *There is an assumption that when teachers transmit feedback information to students these messages are easily decoded and translated into action. Yet, there is strong evidence that feedback messages are invariably complex and difficult to decipher, and that students require opportunities to construct actively an understanding of them (e.g. through discussion) before they can be used to regulate performance. (Nicol & Macfarlane-Dick, 2006)*

One simple strategy to check whether students understand is to have them repeat back to you what their next steps are. This brief step of asking the student to rephrase and repeat what you want them to do will quickly uncover whether or not the next steps are clear to the student.

It's important to note that this recommendation does not mean you have to avoid academic language; rather, you can help students start to incorporate it into their own vocabulary by modeling the use of the language, and encouraging students to use it as well. You may need to take the time to clarify the vocabulary as part of ensuring that students understand their feedback.

To illustrate this idea, let's return to the example of Anna using ratio tables. Here's her teacher's formative feedback to her:

You're correct that proportional relationships can be shown in tables with an input and output, and the numbers go up by the same amount in each column. But remember that we talked about how proportional relationships have a constant factor—a number you can multiply each input by to get the output number—and that factor doesn't change. I don't see that constant factor in the table you've created. See if you can revise your table using this idea of a constant factor.

Anna's teacher used the terms *proportional relationship*, *input* and *output*, and *constant factor*. Depending on what the teacher knows about Anna's knowledge of those terms, he or she might add in some quick checks to be sure Anna understands the feedback:

Teacher: You're correct that proportional relationships can be shown in tables with an input and output, and the numbers go up by the same amount in each column. But remember that we talked about how proportional relationships have a constant factor—a number you can multiply each input number by to get the output number—and that factor doesn't change. I don't see that constant factor in the table you've created. Can you remind me what a factor is?	Anna: It's a number that you multiply by another number to get an answer. Um . . . and it's a number that goes into another number evenly.
Teacher: And what does it mean for something to be constant?	Anna: It stays the same.
Teacher: So if I gave you this table *(writes one down quickly)*, could you tell me what the constant factor would be in this table?	(Anna looks for a moment, then identifies a constant factor of 3.)
Teacher: OK, so see if you can revise your table using this idea of a constant factor.	

A Note on a Common Pitfall: The Dangers of Hit-and-Run Feedback

A potential and common pitfall is to cut short the opportunity to check that students understand the feedback, resulting in what we call *hit-and-run feedback*. This can be due to a sense that not enough time is available or to not wanting to get pulled into a prolonged interaction with a student. This pitfall showed up when several teachers described to us a dilemma they were trying to sort out regarding how much feedback to provide to students. They were trying to move from providing too much information to students to trying to provide a useful hint, cue, or model. They felt that sometimes when they tried to give students less information, the students would then ply them for more information, and they could easily revert to providing too much information. Their solution was to give students a hint and immediately move away so students couldn't keep digging for more information. In an effort to extract themselves from a prolonged interaction with a student digging for more help on how to solve a problem, they were leaving out the important final step of making sure that students understood the feedback they'd received and were clear on next steps for using that feedback.

> **Recommendation 5: Follow up to find out what the learner did with the feedback they received.**

All teachers are pretty accustomed to giving students assignments and having some kind of accountability for the completion of those assignments. This recommendation is merely an extension of that habit, applied to students' actions on the feedback they've received. Perhaps this recommendation could be viewed as just the flipside of a common pitfall: *failing* to follow up to find out what the learner did with the feedback they received. Once students have received feedback, it can be tempting to leave it at that and hope that the students will digest the feedback somehow and that it will then have at least some minor impact on their future work and thinking. However, without acting on the feedback, the feedback may have little or no impact. So it is valuable not only to build in opportunities for students to act on feedback but also to build in ways to make students accountable for acting on the feedback they received. This could take a variety of forms, such as having students

- make revisions to their work and turning in the revisions the next day;
- meet with a partner to give each other feedback, then, after making revisions, meeting with that same partner again to trade work and determine if the partner's feedback was addressed adequately; or
- write about what they learned from the feedback that they can use in future problem situations, or having them reflect on how they used the feedback and how it did or did not help them.

Classroom Material: Go to Resources.Corwin.com/ CreightonMathFormativeAssessment to access this resource.

Feedback Templates. This printable file provides templates that include "Feedback to Feed-Forward," which students can use to reflect on how feedback might help them in the future.

Checking in also provides you with the opportunity to see how useful your feedback was for the student. As you are practicing the new skill of providing formative feedback, you can monitor how your skill is progressing by how often and how well your students' learning has advanced. You can elicit more evidence to see whether they learned what the feedback intended to help them with—and if not, you have an opportunity to troubleshoot why not.

The Student's Role and How You Can Develop and Support It

The student has a critical role in the formative assessment process in both understanding and responding to the feedback he or she receives. Returning for a moment to Thinking Like a Self-Regulating Learner (Figure 4.4), we see that the student has to use the feedback, first to understand where he or she is currently in relation to the learning intention and then, if it is not yet reached, to understand the necessary next steps to take. The student then gets closure by acting on the feedback.

Figure 4.4   Feedback in Self-Regulation

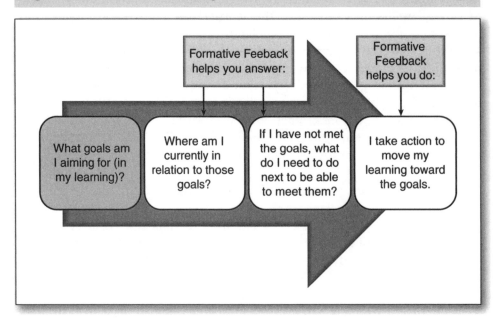

While some students are able to figure out the answers to these self-regulation questions, many students are not, and they need support to learn how to do so. We've seen that the first part of the teacher's role in formative feedback is to provide the feedback to students about their learning, and the second key part—teaching students how to *understand* and then *respond to* the feedback they receive—is equally important. Throughout this section, we illustrate ways that teachers can do this.

TEACHERS' VOICES

We model everything from behavior in the hallway to getting lunch in the cafeteria to our own classroom routines. We know students need to be explicitly told how to perform a routine or process, and yet we expect them to know what to do with all of the comments we give them without any explicit modeling. I realized I needed to pay attention to this.

Understanding

Understanding feedback has two aspects to it. Because students will have experienced different kinds of feedback, students need to understand what you're setting out to do when you give *formative* feedback. They also need to understand specific feedback when a teacher, or a peer, provides it. The following recommendations address both aspects.

Recommendation 6: Explicitly teach students about the meaning, purpose, and use of formative feedback.

Taking the time to teach students about formative feedback has several different payoffs for both you and your students. First, it helps your students start to learn how to use the success criteria to provide feedback to themselves. As we learn something, we all have an ongoing stream of feedback in our heads that we give ourselves, telling ourselves whether we're being successful at something. For example, as I learn to do rolls on a snare drum, I can see it is harder to make the proper movements with my left hand than with my right, and therefore, I tell myself I need to spend more practice time focusing on my left hand. Students provide themselves with the same kind of internal feedback in math class, though that internal feedback can end up being pretty nonspecific ("You're not very good at math . . . ") and not at all linked to any learning (". . . so you better try hard today to not get called on!"). Focusing such internal feedback on the specifics of what they're learning in any given mathematics lesson (i.e. the learning intention and success criteria) is an important part of what self-regulation is about. As they gradually start learning to self-assess against the success criteria, they can learn to focus on their successes, not just their challenges, and they can begin to gain more confidence in their ability to "do math."

Second, when students understand the purpose behind the feedback you provide to them, they can better trust that the feedback is meant to help them take specific next steps so they will better understand what to do with it. They will understand when you identify the things they are doing well, so that they will know to continue with that thinking, and they will understand when you identify an area of improvement that this is where they can focus their energies to get closer to what you want them to learn.

Taking the time to teach students about feedback is an important part of helping students learn to become more self-regulating. One of the first things students need to understand is what *is* formative feedback? In response to this need, teachers in our professional development program asked for lessons to guide this introduction, so we worked with them to develop a lesson plan that introduces students to formative feedback; this lesson is available on our website, and we point it out again in Chapter 5, "Developing Student Ownership and Involvement in Your Students."

Classroom Resource: Go to Resources.Corwin.com/ CreightonMathFormativeAssessment to access this resource.

Introduction to Formative Feedback (for Students). This printable files provides a lesson plan to introduce your students to formative feedback.

Classroom Material: Go to Resources.Corwin.com/ CreightonMathFormativeAssessment to access this resource.

Feedback Poster. This large-format PDF can be posted in your classroom for ongoing reference to the characteristics of feedback. It is suggested for use with the "Introduction to Formative Feedback (for Students)" lesson. The file can be printed poster-sized through your local office supply or copy center.

As many people have experienced, it is difficult to really learn something when you are only hearing about it or reading about it; deeper learning occurs when you actually try to do it. Providing feedback to peers is a good learning exercise for both the giver and the recipient, so when the teachers we worked with asked, "What comes next?" we worked with them to develop a second follow-up lesson to focus on helping students learn to provide useful formative feedback to peers. Teaching students how to give feedback calls for them to apply what they know about formative feedback to the creation of actual examples of feedback in their own words. As they practice articulating and giving feedback, it solidifies their understanding of what formative feedback is. Providing feedback also helps students with their own understanding of the mathematics content. We also talk about that more in Chapter 5, "Developing Student Ownership and Involvement in Your Students."

Classroom Resource: Go to Resources.Corwin.com/ CreightonMathFormativeAssessment to access this resource.

 Peer Feedback. Resources for supporting your students in providing peer feedback are included in Chapter 5; however, you may want to take a look now. Look for Chapter 5 Resources. This lesson introduces providing feedback to peers.

Once you have been explicit about what formative feedback is, and students have had time to practice providing feedback, providing additional opportunities on an ongoing basis is important for continuing to build students' abilities to provide feedback. These opportunities can be provided in multiple ways including using structured templates so all students participate in the giving and receiving of feedback and using feedback as the focus of class discussions. Examples of feedback-focused class discussion include showing an example of student work and asking pairs to discuss feedback or asking for two or three volunteers to provide responses to (1) what has been met, (2) what hasn't been met, and (3) a hint, a model, or a cue for next steps. For the option using volunteers, even though not all students are providing the feedback, all students have the opportunity to hear feedback and think about whether the feedback meets the characteristics.

Classroom Material: Go to Resources.Corwin.com/ CreightonMathFormativeAssessment to access this resource.

 Feedback Templates. This printable file provides templates that can help students give and use feedback.

Responding

The other part of the students' role with regard to formative feedback is to respond to the feedback they receive. In order to respond to it, students need both opportunity and motivation to do so.

> **Recommendation 7: Provide time and support to help students be successful in acting on the feedback.**

Providing time for students to respond to feedback may be one of the most overlooked elements of formative assessment practices, yet it is so necessary. If the students don't have an opportunity to respond to feedback, there is no way for a teacher to determine whether the feedback has been of any use in helping students adjust their learning. Of course, teachers rarely feel like they have enough time to do all the things they need to do in the classroom, and providing these opportunities often seems like yet another thing that competes for time. However, teachers can distribute the time needed for responding to feedback across class time, another structured time (such as study periods, learning labs, after-school sessions), and time at home (as homework).

> ### TEACHERS' VOICES
>
> I have learned that providing students time to act on their feedback is essential to formative assessment. Students need to understand the value of feedback and how feedback moves learning forward. It also teaches them to reflect on their work, understanding it is not (never!) a finished product.

Having the opportunity to respond to feedback not only involves the time needed but also a clear understanding of how to go about doing so. One of the valuable things a teacher can do is to model for students how to respond to feedback. For example, using anonymous student work, teachers can provide feedback on the work and then model how to process the feedback and to determine next steps. This modeling might include a think-aloud in which the teacher (1) reviews the success criteria to understand where the feedback came from, (2) restates the next steps, (3) carries out the next steps, and (4) checks the new results against the success criteria. After providing such a model a few times, the teacher can again provide anonymous student work with feedback and have students discuss in pairs or small groups what to do with the feedback provided. By using work samples and discussing how to use the feedback, students will be better able to determine how to use the feedback they receive individually.

Teachers can also create various routines for students around what to do with feedback. For example, recall Mr. Patari's strategy on page 125 of posting examples of feedback he was giving to students as they worked on a task, then having the class review the feedback together and determine which feedback applied to their own work. This could initially be a routine that he does frequently during class that he could later modify to encourage students to become more independent in their follow-up on feedback, as in this example:

> *Over time, as students got more comfortable with Mr. Patari's process, he started to modify it. As he recorded the feedback on the board, he also recorded the names of students who received particular feedback and were able to use it to successfully revise their work. The next time a different student needed similar feedback, instead of providing the feedback himself, Mr. Patari referred the student to his or her classmate who had used the feedback successfully, in order to discuss next steps.*

 Classroom Resource: Go to Resources.Corwin.com/ CreightonMathFormativeAssessment to access this resource.

 Feedback Response Strategy. This printable file describes a strategy to support students' efforts to act on formative feedback.

■ CONCLUSION

Giving effective feedback to students sounds simple, but it is remarkably complicated to do well. In this chapter, we focused on the various characteristics of formative feedback and how those characteristics have an impact on how students take in and act on the feedback, whether it is provided to an individual, a small group, or a whole class. We've also discussed the importance of your role, both in providing formative feedback to your students and in supporting them as they learn how to understand and use the feedback they receive. When you have provided effective formative feedback to a student, you have given him or her enough information to move his or her own learning forward without taking the intellectual work out of the student's hands—or mind. Letting the student know what he or she has done well is as important as letting him or her know what needs to be done to fix it or where more work is needed. Then providing an opportunity to use feedback to revise the work and learning is a vital step to gaining the full benefit of formative feedback and bolsters the student's belief that he or she can be successful. Thus, time to provide or receive and then to act on formative feedback is necessary if the learning is to stick, and it is time well spent for teacher and students alike.

Formative feedback is based on a judgment call on the part of the teacher of what constitutes too much or too little for a particular student—as with all things in the classroom, it depends very much on knowing your students and gauging what's right for them. It is informed by your knowledge of what your class has been doing and where the students are heading in their learning. The very same feedback can be highly effective for one student, yet completely ineffective for another, because feedback is responsive to what an individual student needs as evidenced by his or her thinking and understanding. In the end, we encourage you, our reader, to practice as often as possible, keeping the characteristics of formative feedback in mind and making your best effort. Over time, you will undoubtedly hone your skill!

Recommendations in This Chapter

1. Include planning for formative feedback as part of your lesson planning.

2. Provide feedback to students that points out what criteria have been met (if any), what criteria have not been met (if any), and a suggestion for a next step to move learning forward.

3. Keep the feedback specific and focused on the success criteria.

4. Make sure students understand the feedback they're receiving.

5. Follow up to find out what the learner did with the feedback they received.

6. Explicitly teach students about the meaning, purpose, and use of formative feedback.

7. Provide time and support to help students be successful in acting on the feedback.

Reference Resources: Go to Resources.Corwin.com/ CreightonMathFormativeAssessment to access these resources.

Chapter 4 Summary Cards. This printable file provides Summary Cards that include recommendations in this chapter.

Formative Assessment Recommendations. This printable file provides a summary of all the recommendations in this book in a printable (letter size) format.

RESOURCES ■

The following sections present some resources that can help you implement the recommendations. Each of these resources is referenced in earlier sections of the chapter, but here we provide a consolidated list. All resources can be found at Resources.Corwin.com/CreightonMath FormativeAssessment.

Learning Resources

These resources support your learning about formative assessment practices related to evidence gathering and interpreting:

- **Feedback Characteristics—Content** provides practice deciding whether feedback examples meet the characteristics related to content.
- **Feedback Characteristics—Usability** provides practice deciding whether feedback examples meet the characteristics related to student usability.

Reference Resources

These resources summarize key ideas about formative assessment practices related to evidence gathering and interpreting:

- **Chapter 4 Summary Cards** provides index-sized Summary Cards for the following:
  - **Characteristics of Formative Feedback** summarizes the characteristics described in this chapter.
  - **Recommendations for Formative Feedback** provides the recommendations from this chapter.
- **Formative Assessment Recommendations** is a printable PDF that includes all the recommendations from this book.

Planning Resources

This planning tool supports your lesson planning when integrating formative assessment into your mathematics lessons:

- **Formative Assessment Planner Templates** can be used for creating a formative assessment plan for a lesson. One version includes a portion in which you can record model responses for the success criteria.

Classroom Resources

These resources illustrate various classroom routines that you can use during instruction when gathering and interpreting evidence with your students. Each routine provides a structure that you can use or adapt to routinize your practice around the use of evidence.

- **Framework for Using LI and SC With Students** describes a way to help students understand and use learning intentions (LI) and success criteria (SC) during a lesson. It describes actions for both teacher and students to take at the start of the lesson, at midway points during the lesson, and at the end of the lesson.

- **Revisiting Strategies** describes several strategies for use when revisiting LI and SC throughout the lesson, including a strategy for providing feedback. More about these strategies is included in Chapter 5.
- **Teacher Summary Cards for the Strategies** are large-sized cards that provide step-by-step summaries of *teacher* moves for the strategies described in Revisiting Strategies.
- **Introduction to Formative Feedback (for Students)** is a lesson plan for introducing the characteristics of formative feedback to students.
- **Peer Feedback** is a lesson plan for introducing and practicing how to give formative feedback to peers.
- **Feedback Techniques** lists formative assessment classroom techniques (FACTS) related to feedback that can be found in the book *Mathematics Formative Assessment: 75 Practical Strategies for Linking Assessment, Instruction, and Learning* by Keeley and Tobey (2011):
  - Collaborative Clued Corrections
  - Comments-Only Marking
  - Feedback to Feed-Forward
  - Peer-to-Peer Focused Feedback
  - Partner Speaks
  - Two Stars and a Wish
- **Feedback Response Strategy** describes a strategy that you can give your students to help them respond to feedback.

Classroom Materials

These resources can be copied or re-created to be used in your classroom.

- **Student Summary Card for the Strategies** are large-sized cards that provide step-by-step summaries of *student* moves for the strategies described in Revisiting Strategies. This can be distributed to each student or pairs of students for use when learning about a strategy.
- **Feedback Templates** for using the feedback techniques described earlier:
  - Providing Feedback includes Feedback Sandwich and Peer to Peer Focused Feedback
  - Reflecting on Use of Feedback includes Feedback to Feed-Forward and Planning from Feedback
- **Feedback Poster** can be posted in your classroom for ongoing reference after introducing students to the characteristics of feedback. The file can be printed poster-sized though your local office supply or copy center.

5

Developing Student Ownership and Involvement in Your Students

You have now looked in depth at the nature and purpose of three of the four critical aspects of formative assessment: (1) learning intentions and success criteria, (2) eliciting and interpreting evidence of students' learning, and (3) formative feedback. Now that we've discussed the interdependence of these three critical aspects, we can delve more fully into the fourth and most important critical aspect of formative assessment: developing student ownership and involvement.

In the previous chapters, we've touched on some elements of the students' role in each of the various critical aspects. In this chapter, we spell out more specifically what that role is for your students within a mathematics classroom using formative assessment practices, what they need to learn in order to step into this role, and what you can do to support them and help them learn. We also summarize the various planning resources and classroom resources available to you to help you support your students in becoming more self-regulating, and we discuss in more depth how you can use those resources to best effect.

■ STUDENT OWNERSHIP AND INVOLVEMENT

Developing students' ownership and involvement in their mathematics learning is about helping students learn to internalize the three key questions and fourth action step for self-regulation (Figure 5.1). It means helping them learn what they need to *pay attention to* when thinking about their own learning and what they need to *do* so that they can answer the three key questions and take steps to move their learning forward.

Figure 5.1   Student Ownership and Involvement in Self-Regulation

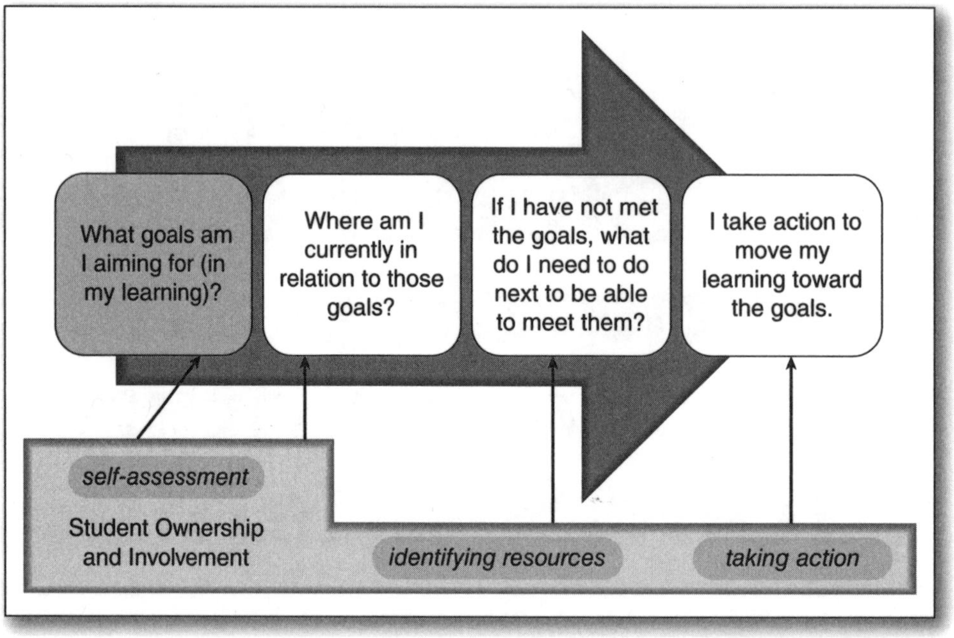

What goals am I aiming for (in my learning)?

Where am I currently in relation to those goals?

If I have not met the goals, what do I need to do next to be able to meet them?

I take action to move my learning toward the goals.

self-assessment

Student Ownership and Involvement

identifying resources

taking action

Recall that our definition of a student who is *self-regulating* is one who is able to effectively engage in the process laid out in Figure 5.1 in its entirety: to answer the three questions shown in the diagram and to then take action to move his or her learning forward. Self-regulating students monitor and self-assess their own learning against success criteria. They are more likely to be engaged and involved in part because the goal of their learning is clear to them and the criteria for judging their success in reaching that goal is also clear to them. And they have a sense of ownership over their own learning because they have a clear picture of what their role is as a learner, and they have learned some skills that allow them to step into that role.

As we've discussed self-regulation in earlier chapters, we've already presented an initial picture of what a self-regulating student looks like. In this chapter, we be honing our definition of *student ownership and involvement*, and so looking at the language we be using is worthwhile. Let's start from some common ways that teachers often talk about this:

I want students to take responsibility for their own learning.

Students need to take ownership of what they're learning.

I want my students to be involved in their own learning.

Words like *involvement, responsibility, engagement, ownership, monitoring, self-assessment,* and *independent learners* often are used to try to capture a somewhat elusive picture of students who have developed strong meta-cognitive skills and are confident and proactive in regard to their own learning. Yet, these terms may not reflect all that a self-regulating student is, or does:

- Students can be *engaged* when they are interested and paying attention, and sometimes also *involved* when they are participating, without necessarily being able to do any *self-assessment*—evaluating how their own work measures up against some criteria.
- Students can *monitor* their own work as they determine whether they are learning in an effective way, but if they realize they're having difficulty with an idea, they may not necessarily know what to do about it.
- Students can take *responsibility* or show *independence* when they are proactive about completing their work or about seeking out help when they need it but still not have any sense of *ownership* of the learning process if they defer to the teacher for all decisions about their learning.

In this chapter, we use the phrase *student ownership and involvement* to encompass the entirety of the process of being a self-regulating student.

One of the goals of developing student ownership and involvement in mathematics learning is to start to shift students' internal dialogue away

Table 5.1   Summary of Internal Dialogue Examples

Level of Self-Regulation	What goals am I aiming for in my learning?	Where am I currently in relation to those goals?	If I have not met the goals, what do I need to do next to be able to meet them?	I take action to move my learning toward the goals.
Least Self-Regulating	*Do the work to my teacher's satisfaction.*	*I don't know; my teacher will tell me.*	*I will wait for the teacher to help me.*	*Only if the teacher will tell me what to do next, or when to do it, or how to do it.*
In Between Least and Most Self-Regulating	*Learn some math ideas (may not be sure what they are?)*	*I'll evaluate based on right answers and whether I feel like I get it.*	*Feedback helps me correct wrong answers; then, just try harder.*	*I should use my resources; my teacher will point me toward the right ones.*
Most Self-Regulating	*My job is to learn something new—the learning goal for today's lesson.*	*I can try to check myself against the criteria for good work. I can also look at any feedback I've received about my work or my thinking.*	*I may have some ideas of my own. If I don't, or if I'm unclear, I'll talk to my teacher or peers to figure out how to improve my work or adjust my thinking.*	*I can get myself started. If I get stuck, I can refer to my resources or the feedback I got. I can also get help from a peer or my teacher.*

from absolutes, such as "I can't do math," and toward teaching students how to understand and pay attention to specific criteria for what success in today's math lesson looks like and how to use feedback on their work to help them successfully produce work that meets the criteria and therefore the learning intention. Let's look at what might be the typical internal dialogue going on for students who are in different places along a spectrum of levels of self-regulation, as we started to in Chapter 1. Table 5.1 (p. 155) compares some representative internal dialogue for each part of self-regulation.

TEACHERS' VOICES

Each year I have some students who enter my class turned off to math. They *expect* to be wrong and see no point in wondering why they are. The beauty of formative assessment is in slowly peeling away the layers of reluctance. By referring to the learning intentions and success criteria, it inevitably becomes apparent to these students that often their problems aren't in understanding the new math concept but may more likely be something simple like a fact error. Once they can celebrate the pieces they've grown to understand, they are more willing to take the next step: caring enough to take action on the feedback because it seems manageable, doable. Watching the transformation in their self-image as mathematicians is so rewarding! As soon as the focus in feedback switches to what they did well, the walls start to come down and they start becoming vested. Just last week after our state testing, one such student who made huge gains since fall proudly told me that he was beginning to believe me when I said he was a natural mathematician. Those are the moments that a teacher lives for!

This student self-regulation process also appears in the Formative Assessment Cycle. In the During the Lesson portion of the Formative Assessment Cycle, each of the half-circles describing the teacher role is mirrored by a corresponding half-circle that describes elements of the students' role. (See Figure 5.2.) These half-circles describe the students' role in eliciting evidence, interpreting evidence, and responding to feedback. (As you will see in Chapter 7, "Establishing a Classroom Environment," students also have a role in supporting formative assessment that does not yet appear in our evolving Formative Assessment Cycle.)

■ WHAT DO STUDENTS NEED TO LEARN?

In order to develop a sense of ownership and involvement in their mathematics learning, students need both to learn a specific set of actions to take in math class, as well as to develop a can-do mind-set, illustrated in the teacher quote above, that allows them to take these actions proactively. In our experience, this is a bit of a chicken-or-egg situation in terms of which comes first, the actions or the mind-set; the development of one seems to bolster the development of the other. We use the self-regulation diagram to talk about both. For each question within the self-regulating diagram, we put forth a set of statements about what students need to learn. We

Figure 5.2 Student Ownership and Involvement in the Evolving Formative Assessment Cycle

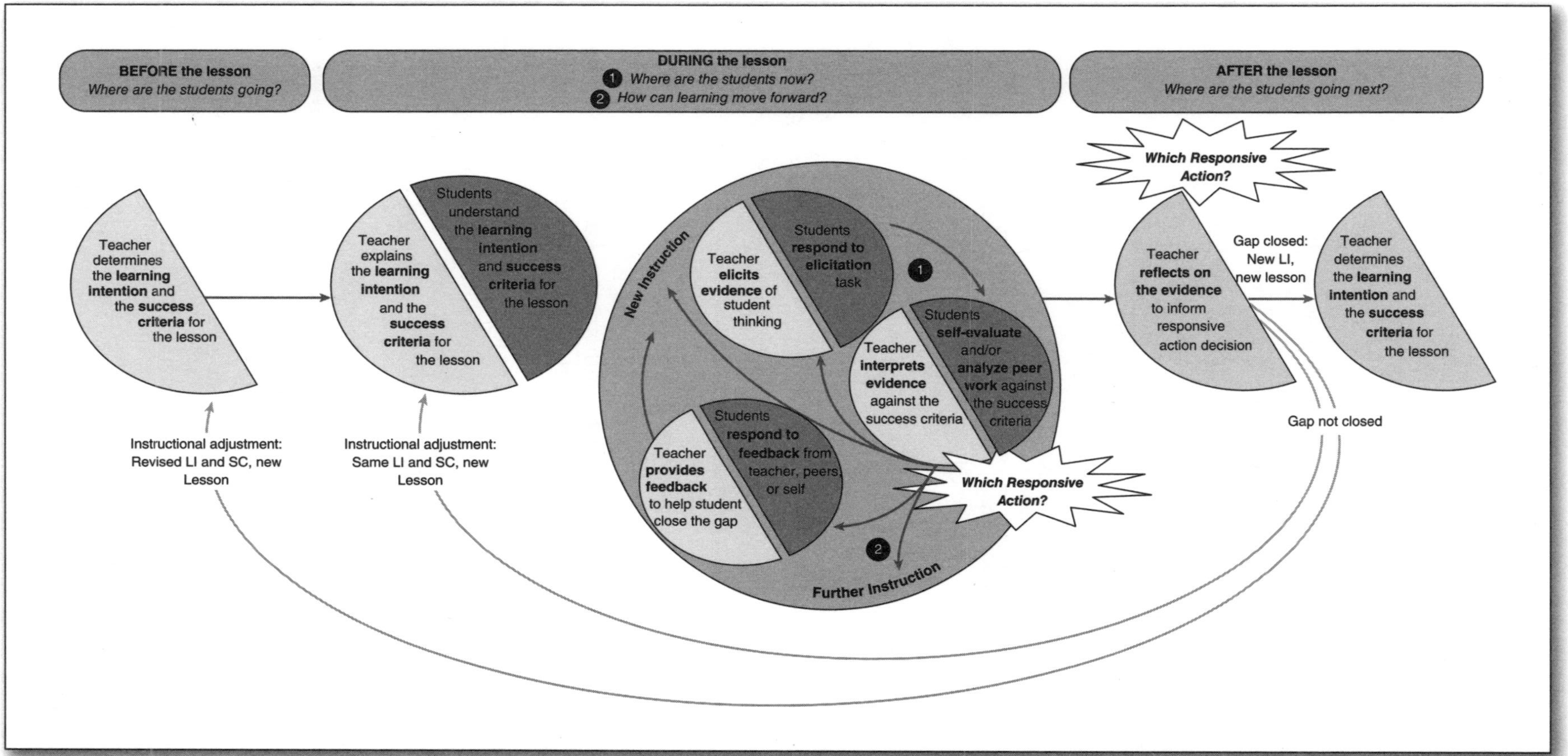

BEFORE the lesson
Where are the students going?

DURING the lesson
1 *Where are the students now?*
2 *How can learning move forward?*

AFTER the lesson
Where are the students going next?

Teacher determines the learning intention and the **success criteria** for the lesson

Instructional adjustment: Revised LI and SC, new Lesson

Teacher explains the **learning intention** and the **success criteria** for the lesson

Students understand the learning intention and success criteria for the lesson

Instructional adjustment: Same LI and SC, new Lesson

New Instruction

Teacher **elicits** evidence of student thinking

Students respond to elicitation task

Teacher **interprets evidence** against the success criteria

1

Students self-evaluate and/or analyze peer work against the success criteria

Which Responsive Action?

2

Teacher **provides feedback** to help student close the gap

Students respond to feedback from teacher, peers or self

Further Instruction

Which Responsive Action?

Teacher **reflects on the evidence** to inform responsive action decision

Gap closed: New LI, new lesson

Gap not closed

Teacher determines the learning intention and the **success criteria** for the lesson

157

then follow each of those statements with a more in-depth look at why students need to learn the idea and considerations about how you can help them learn.

The First Question: Learning to Use Learning Intentions and Success Criteria to Clarify Goals

The first self-regulation question is, "What goals am I aiming for in my learning?" as shown in Figure 5.3.

Figure 5.3 The First Question

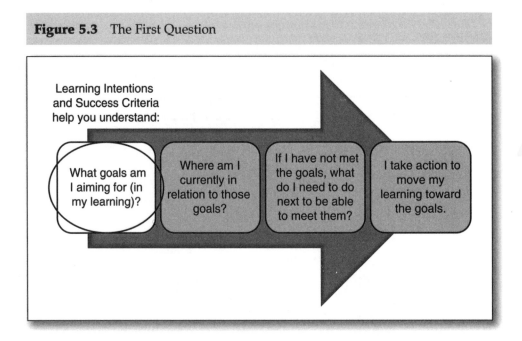

In Chapter 2, "Using Mathematics Learning Intentions and Success Criteria," we said that a self-regulating student responds to this question by focusing on reaching the learning intention for the current lesson. To get that focus, students need to learn *how to make use* of learning intentions and success criteria. Many students are accustomed to seeing some kind of mastery objective, learning target, or "I can . . ." statement provided by their teacher, so having a learning intention that states the goal for their learning in a lesson and some criteria that describe that learning may not seem all that different to them. The difference for students comes in learning both how *you* are using them as a teacher and how *they* can make use of them to further their own mathematics learning. Specifically, students need to learn

- how you will be using learning intentions and success criteria in lessons, so they will understand that they should pay attention to them as well;
- to make an effort to understand what the learning intention and success criteria mean when you introduce them, as well as throughout the lesson; and

- to talk about their mathematical reasoning, not just their answers, so that you, and they, can accurately uncover how they are making sense of the mathematics.

Let's look more closely at each of these needs.

- **Students need to learn how you will be using learning intentions and success criteria in lessons, so they will understand that they should pay attention to them as well.**

   Initially in your implementation of formative assessment practices, many of your students may not pay much attention to the learning goal beyond noting it at the start of class as some indication of the direction of the current lesson. However, you can quickly start to help them learn that paying attention to the success criteria, and thus the learning intention, is helpful to them. As you use success criteria as a way to help them understand what reaching the learning intention will look like, as guidelines for seeking and interpreting evidence of their mathematics learning, and as a basis for providing feedback, students begin to see that the success criteria are some useful guideposts they can use, too.

TEACHERS' VOICES

I have become more aware of how learning intentions and success criteria can help both me and my students. I recently had a revelation about this. Writing good learning intentions and success criteria *first* and planning activities *second* was one change I had to make. It makes it so much clearer for both me and my students what we should be focusing on. It is easier for me and my students to be on the same page.

- **Students need to learn to make an effort to understand what the learning intention and success criteria mean when you introduce them, as well as throughout the lesson.**

   It's not enough for students to just sit and listen to you or a classmate read a learning intention to them. The goal is for students to think about what the learning intention and success criteria mean, to determine whether they understand what is expected that they learn, and to speak up if they don't. Students will be motivated toward this goal when they begin to see the learning intention and success criteria as useful, and that takes time to develop. As students gradually see how learning intentions and success criteria are helpful in providing focus for their learning, in evaluating their work, and in receiving and acting on useful feedback that feels manageable to them, they will begin to tune in to learning intentions and success criteria more closely.

- **Students need to learn to use the learning intention and success criteria as reference points when they are unsure about where to focus their attention during instruction.**

Sometimes there are situations in math class when many different math skills or math concepts all seem to be vying for attention at the same time. For example, when Mr. Benton's eighth-grade algebra class was discussing the slope of a line as a measure of rate of change, the class talked about the idea of slope, ways to calculate the slope, and how the slope was different than the y-intercept in a linear equation. They also reviewed how to appropriately plot points on a graph and the difference between positive and negative slope. Although all these concepts and skills are closely interrelated, some students can get overwhelmed trying to keep track of what all these mathematical concepts and skills are and how they are related. Given clear learning intentions and success criteria, students can learn how to pause, to look back at the learning intention and success criteria as a reminder of what the primary focus is for the lesson, and then to use the learning intention and success criteria to get clarity on what are most important ideas to glean for the day's lesson. And this checking-back can happen at any time throughout a lesson. Later in this chapter, we review how you can use some of the resources we've provided to model this for students.

The Second Question: Learning to Use Success Criteria, Evidence, and Formative Feedback to Self-Assess

The next self-regulation question is, "Where am I currently in relation to those goals?" as shown in Figure 5.4.

Figure 5.4 The Second Question

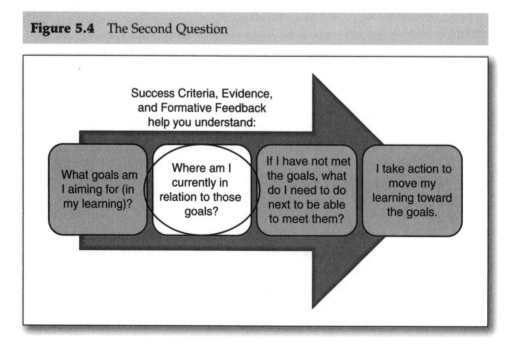

Looking back at our description of the most self-regulating student, we see that self-regulating students respond to this question by using criteria for good work as guidance for evaluating their work. In a classroom using formative assessment practices, these are the success criteria. Self-regulating students also draw on feedback from teacher or peers to help determine the extent to which their work meets the success criteria. To help them do that, students need to learn what constitutes evidence of their mathematical learning, *beyond just their answers to math problems*, and the role that evidence can play in determining how well they are doing. They also need to understand how feedback can provide useful information about their learning. Specifically, students need to learn

- to talk about their mathematical reasoning, not just their answers, so that you, and they, can accurately uncover how they are making sense of the mathematics; and
- to use the success criteria to self-assess their work in terms of what ways and to what extent they've met the success criteria.

Let's look at each of these needs.

- **Students need to learn to talk about their mathematical reasoning, not just their answers, so that you, and they, can accurately uncover how they are making sense of the mathematics.**

Students are accustomed to being called on to provide correct answers, but they may be less accustomed to having conversations in math class about their reasoning. Many students may think that what you are asking them to provide when you say "explain your thinking" is a laundry list of steps that they followed to solve a problem: "First, I found a common denominator, then I adjusted the numerators, then I added the numerators but left the denominators the same." As students learn to engage in conversations about their thinking, they begin to focus on questions:

  o What were you paying attention to in the problem?
  o *Why* did you choose to do that? (For example, why did you add the numerators but leave the denominators the same?)
  o What did you do next when you were stuck, and why?
  o What connections are you making between mathematical ideas that you know?

With these kinds of questions, the conversations move away from a focus on *what* the student did, and toward a focus on *how* and *why* the student was thinking in a particular way.

- **Students need to learn to use the success criteria to self-assess their work in terms of what ways and to what extent they've met the success criteria.**

Student self-assessment is a key to answering "Where am I in relation to the goals?" Students need to learn to gauge their work in relation to the success criteria and to pay as much attention to *how* or *to what extent* they've met the criteria as to *whether* they've met the criteria. Initially, students tend to see the number of right answers as the best indication of whether they are meeting the criteria. As they learn to use the success criteria in their self-assessment, their self-assessment starts to shift from a general "I get it/I don't get it" to more specific identifications like, "I understand this part but still don't understand this other part," "I think I just need more practice," "I really don't understand this yet," or "I understand this well enough to explain it to someone else; I feel ready to move on."

TEACHERS' VOICES

Students are more specific with the questions they ask of me and of each other when they are stuck than they used to be. When students are working in pairs, they compare their approaches to a problem, and I hear them say things like, "I have the same thing as you until this point. Then our work is different. I get this part, but how did you get this?"

An important part of this shift involves learning to distinguish between their actual mastery of any particular content and their perception of how well they understand it. Early in their work with self-assessment, many of the middle school students we observed would readily assess themselves as having learned a topic when both we, and their teacher, knew they had not. What they had not yet learned was the difference between "how I think I'm doing" and "how I'm actually doing" when they solve problems and explain what they know. For many of them, "how I think I'm doing" often equated with how easily they could come up with a solution method (whether or not it was correct) or how confused or not they were by a particular math problem—it had little or nothing to do with the success criteria. Over time, with practice, they started to see the difference and to use the success criteria as guidance for their self-assessment.

 Classroom Resource: Go to Resources.Corwin.com/ CreightonMathFormativeAssessment to access this resource.

 Self-Assessment Techniques. This printable file provides a list of additional strategies for student self-assessment that can be found in the book *Mathematics Formative Assessment: 75 Practical Strategies for Linking Assessment, Instruction, and Learning* by Keeley and Tobey (2011). A brief description of each strategy is included.

The Third Question: Learning to Determine Next Steps

The third and final self-regulation question is, "If I have not yet met the goal, what do I need to do next to be able to meet it?" as shown in Figure 5.5.

Figure 5.5 The Third Question

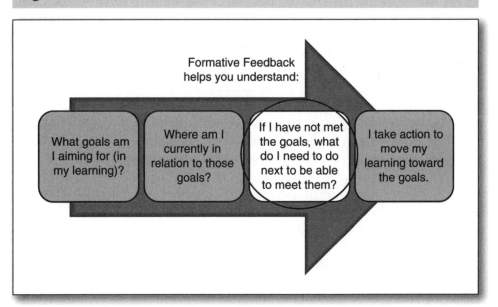

Looking back at our description of the most self-regulating student, we see that self-regulating students respond to this question by sometimes making their own determination of what they need next, which may or may not include seeking formative feedback from a peer or a teacher. To do that, students need to learn ways to move their learning forward without getting formative feedback and how to use any formative feedback they do receive. Specifically, students need to learn

- what formative feedback is and how to use it to support the self-assessment process;
- to use various resources (such as their notes, different problem-solving strategies, and textbooks, articles, and tutorials on the Internet) to move their learning forward;
- to use peers as well as their teacher as resources to help them when they are stuck; and
- to look for the part of the feedback they receive that has hints, cues, or models that suggest productive next steps.

Let's look at each of these needs.

- **Students need to learn what formative feedback is and how to use it to support the self-assessment process.**

  Formative feedback provides some of the information students need in order to determine how they are doing in relation to reaching the learning intention, but students need to trust that the feedback will help them learn if they use it properly; otherwise, they are likely to ignore it or make a halfhearted attempt just to appease the teacher. To gain this trust, students need to understand how the feedback provided helps them see what was met and what was not met, and then they need to compare that with their own assessment of the work against the criteria. Learning how to use formative feedback takes practice and time, but with that practice and time also comes that sense of trust that the feedback will provide information they can understand and use, and that it will make the next steps attainable. As this sense of trust develops, students begin to pay much more attention to formative feedback and even seek it out.

- **Student need to learn to use various resources (such as their notes, different problem-solving strategies, textbooks, articles, and tutorials on the Internet) to move their learning forward.**

  Consider the kinds of hints, models, or cues you might provide students as part of formative feedback. You might tell them to review an example problem from their notes or textbook or suggest a way that they might approach the problem differently. Students who internalize these kinds of responses may be able to give feedback to themselves, essentially, and to try one of those resources without the need to consult with a peer or teacher.

- **Students need to learn to use peers as well as their teacher as resources to help them when they are stuck.**

  All students are accustomed to looking to their teacher for help and often to their peers as well. There may be a select group of students in a classroom who are viewed as the most useful to approach for help, or students may just rely on their friends whether or not they are in fact helpful. Some teachers try to encourage students to use peers as resources by having guidelines such as "Ask Three Before Me," in which students are expected to ask three other students for help before approaching the teacher. The new learning here for students is about seeking help specific to the success criteria from their peers and increasing their comfort with seeking out help from peers as readily as they seek it from their teacher.

- **Students need to learn to look for the part of the feedback they receive that has hints, cues, or models that suggest productive next steps.**

  The importance of the part of the feedback that is the hint, cue, or model with pointers to productive next steps may be elusive to many students, and it may take time and practice for them to understand how valuable this part can be. One important reason to have students learn to provide formative feedback to peers or to a fictitious other student is that, whether they are ever required to provide feedback to another student (this may not be practical for some students), it helps *them* learn to use the feedback *they* receive. Practicing providing formative feedback can raise their level of awareness about this part of the feedback, which helps them become better able to look for it in the feedback they themselves receive.

The Final Step: Taking Action

The final and all-important step of self-regulation is to "take action to move my learning toward the goal," as shown in Figure 5.6.

Figure 5.6   Taking Action

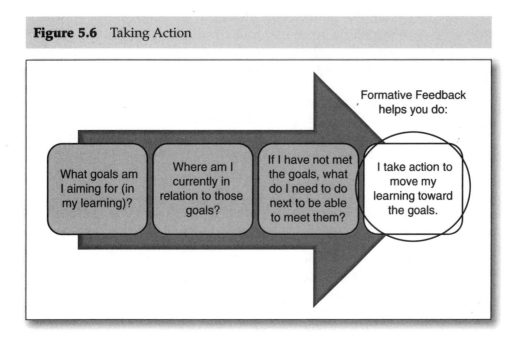

Looking back at our description of the most self-regulating student, we see that the most self-regulating students, after determining a next step such as those indicated above, do not need to rely on the teacher's direction to get started. If they then encounter another barrier, these students have learned to use either the teacher or their peers as resources to help them get moving again. To help them, students need to learn

- to use what they found in a resource (such as their notes, a different problem-solving strategy, a textbook, an article, or a tutorial on the Internet) to take some next steps in their work, and
- to use the hint, cue, or model provided in the feedback to take some next steps in their work.

Let's look at each of these needs.

- **Students need to learn to use what they found in a resource (such as their notes, a different problem-solving strategy, a textbook, an article, or a tutorial on the Internet) to take some next steps in their work.**

  Although knowing what resource to refer to is an important starting point, it is not enough. Students also need to learn how to use the information gleaned from the resource to make changes to their own work.

- **Students need to learn to use the hint, cue, or model provided in the feedback to take some next steps in their work.**

Not only do students need to learn to be the giver of feedback, but they also need to learn to be the receiver of feedback, and that includes following up to take action. You read earlier in the chapter that students need to develop a sense of trust that the feedback they receive will be helpful. After they develop that sense of trust by reviewing the information provided about the extent to which the success criteria were met and considering the hint, model, or cue, the new learning here for students is to then independently take some next steps to act on the feedback provided.

While these steps are summarized in the diagram you've seen several times so far, Figure 5.1, the teachers we worked with asked for a poster that summarized them in a way that would be helpful for their students to see. The resulting Self-Regulation Poster is available on the companion resource website.

 Classroom Material: Go to Resources.Corwin.com/ CreightonMathFormativeAssessment to access this resource.

 Self-Regulation Poster. This large-format PDF summarizes the three self-regulation questions and final step to post in your classroom. The file can be printed poster size through your local office supply or copy center.

■ HELPING STUDENTS DEVELOP OWNERSHIP AND INVOLVEMENT IN THEIR MATHEMATICS LEARNING

There are a collection of specific things you can do to help your students develop a sense of ownership and involvement in their mathematics learning, and the good news is that your students really can develop it. In the words of one of the teachers from our professional development work:

The focus of all of my training in education had been on the teacher finding ways to keep the students engaged in learning. We were taught to be entertaining, relevant to students' lives, fast paced with a variety of presentation styles, sensitive to various learning styles, and aware of needed wait time while still keeping the pace moving to accommodate ever shrinking attention spans in our ever increasingly instant world. This led to the development of some truly amazing and dynamic math lessons! It required a huge amount of creativity and work on my part, and for the most part, it was rewarding to create these lessons. However, as standardized test results showed, it did not apparently lead to higher levels of student engagement—at least not if that engagement was measured by test scores.

Now, in my classroom, the students are the central focus, not me, not the teacher. I no longer think about how I am going to engage my students; I think it is fair to say that they are busy finding ways to engage me in their learning. "I think I get this." "Wait, what do I do if there is a whole number in the equation? Are the rules we found still the same?" "I am

stuck, could you please look at this?" "Am I on the right track so far?"
"Can I tell you what I am doing here in case I am not doing it correctly?"
The nature of my conversations with my students has changed dramatically and for the better.

In this next part of the chapter, we lay out a set of recommendations that encompass the various things that we've learned can help your students move toward greater ownership and involvement. These recommendations all center around the main ideas of being explicit with your students about what you're doing to implement formative assessment and why, modeling for them the behaviors they need to learn, and providing structures that help them practice those behaviors.

Helping Students Learn About Learning Intentions and Success Criteria

Earlier in the chapter, we listed three things that students need to learn in order to be able to use learning intentions and success criteria effectively at various points in the self-regulation process:

- How you will be using learning intentions and success criteria in lessons, so they will understand that they should pay attention to them as well
- To make an effort to understand what the learning intention and success criteria mean when you introduce them, as well as throughout the lesson
- To use the learning intention and success criteria as reference points when they are unsure about where to focus their attention during instruction

The following recommendations and resources can help you support your students' learning of these three points.

> **Recommendation 1: Explicitly talk to students about what the learning intention and success criteria are and how you are using them.**

One way to help students gain better understanding of what the learning intention and success criteria are is to devote some time in class to teach their purpose, and we've developed a lesson that does that. Devoting attention specifically to introducing the learning intention and success criteria lays an important foundation, since they form the basis for most of your other subsequent formative assessment implementation efforts.

 Classroom Resource: Go to Resources.Corwin.com/ CreightonMathFormativeAssessment to access this resource.

 Introducing Students to Learning Intentions and Success Criteria. This printable file provides an introductory lesson to help students understand the purpose of learning intentions and success criteria.

You can also look for opportunities to point out explicitly to students that you are referring back to the learning intention and success criteria during a lesson, and explain why. This can be a very brief moment in your lesson, or it can be a more substantive revisiting, as we talk about later in the chapter. During these moments, it's helpful to remind students that they can always check the success criteria themselves if they need to remind themselves where to focus their learning or need to self-assess their work.

In Chapter 2, "Using Mathematics Learning Intentions and Success Criteria," you read about how the learning intention and success criteria need to remain always available so students can learn to refer to them throughout the lesson. Some teachers accomplish this by posting them in a regular place in the classroom, written large so they are easy to read from any location in the room. Some teachers choose to list them at the top of any key handouts for the lesson. Other teachers choose to have students copy them down into notes for the lesson. Teachers with Smart Boards will include them as a slide in the lesson's presentation, but then they will also post them, include them on a handout, or have students record them in their notes so they are available for students' and teachers' reference throughout the current lesson. Whatever approach works for you, the important thing is to have them available so that both you and the students can refer to them whenever necessary. We look more at posting learning intentions and success criteria in Chapter 7, "Establishing a Classroom Environment."

Reference Resource: Go to Resources.Corwin.com/ CreightonMathFormativeAssessment to access this resource.

 Images of Posted Learning Intentions and Success Criteria. This printable file provides images of the various ways teachers have posted learning intentions and success criteria for reference and use during a lesson. You will read more about posting learning intentions and success criteria in Chapter 7.

Recommendation 2: Use a variety of ways to clarify for students the meaning of the learning intention and success criteria, including near the start of the lesson and during the lesson.

a. Draw students' attention to the learning intention and success criteria initially to clarify any language and discuss any potentially confusing points.

b. Draw students' attention back to the learning intention and success criteria at numerous points throughout the lesson both to help them stay focused on the intended learning for the lesson and to help further clarify the intended learning.

For students to self-regulate, the students must understand what the learning intention and success criteria mean. If the teacher is the only one who understands the criteria, then only the teacher can gauge the nature and extent of the students' learning. In his work on formative assessment,

Dylan Wiliam (2000) noted that in one study, researchers examined the impact of making sure that students understood the success criteria:

> *[The study] suggests that, at least in part, low achievement in schools is exacerbated by students' not understanding what it is they are meant to be doing—an interpretation borne out by the work of Eddie Gray and David Tall (1994). This study, and others like it, shows how important it is to ensure that students understand the criteria against which their work will be assessed. Otherwise we are in danger of producing students who do not understand what is important and what is not.*

The sharing of learning intentions and success criteria with students can become an integral part of the lesson that strengthens students' learning, rather than a quick prologue or add-on to the lesson. The teaching framework introduced in Chapter 2, "Using Mathematics Learning Intentions and Success Criteria," describes an overall structure for first sharing then revisiting the learning intention and success criteria over the course of a lesson. Within this structure, you can use different specific strategies for initially sharing them and for revisiting them; these strategies are listed as we go through the chapter.

For Recommendation 2a, near the start of a lesson, we suggest four Sharing Strategies to help students gain this understanding. (We say "near the start of the lesson" since this does not necessarily have to be the first thing that is addressed in a lesson.) Each Sharing Strategy is illustrated in more detail using the Framework resources available on the companion resource website.

- *Clarification Strategy:* Spend some time reviewing any unfamiliar language for students. Note: In some cases, clarifying new academic language that students will learn during the lesson can be done more effectively later in the lesson rather than at the start.
- *Cover-Up Strategy:* Help students paraphrase the learning intention and success criteria in their own words.
- *3-Read Strategy:* Have students read the learning intention and success criteria multiple times, each time reading for different information.
- *Think/Pair/Share Strategy:* Give students time to "think, pair, share" about their understanding of the learning intention and success criteria. As the class discusses the shared understandings, you can correct or redirect any misunderstandings.

 Classroom Resources: Go to Resources.Corwin.com/ CreightonMathFormativeAssessment to access these resources.

 Framework for Using LI and SC With Students. This printable file describes a way to help students understand and use learning intentions (LI) and success criteria (SC).

 Sharing Strategies. This printable file describes strategies for sharing learning intentions and success criteria near the beginning of a lesson.

 Teacher Summary Cards for the Strategies. This printable file provides large-sized cards that provide step-by-step summaries of teacher moves for the strategies described earlier.

Classroom Material: Go to Resources.Corwin.com/ CreightonMathFormativeAssessment to access this resource.

Student Summary Cards for the Strategies. This printable file provides large-sized cards with step-by-step summaries of student moves for the strategies described earlier. This can be distributed to each student or pairs of students for use when learning about a strategy.

Clarifying the meaning of the success criteria and the learning intention is part of a critical conversation between teacher and students that goes far beyond posting the learning intention and success criteria for the day and reading them at the start of class. As suggested in Recommendation 2b, you need to take time—not just near the start of a lesson but also during the lesson—to clarify, illustrate, and refine what it means to meet the success criteria. Since simply sharing or clarifying the success criteria at the start of a lesson is usually not sufficient for students to fully understand them, build in opportunities to revisit the success criteria during the lesson to further clarify them and to ensure that students are interpreting them properly.

In addition to clarifying the learning intention and the success criteria, revisiting the success criteria in the midst of your lesson, even if only very briefly, can have other benefits for you and your students. For example,

- you can use the time to take stock of what the class understands at that point in the lesson;
- you can provide some whole-class feedback on one or more of the success criteria, based on what you have seen in their work up to that point; or
- students can take a moment to consolidate their learning during the lesson.

This revisiting can happen very organically for many teachers. Some need to remind themselves to return to the success criteria, but over time they find it begins to feel more natural.

We illustrate different examples of revisiting the success criteria using Mr. Carlton's lesson from the previous chapter, for which his learning intention and success criteria are:

LEARNING INTENTION AND SUCCESS CRITERIA

We are learning today how we can use an equation to represent a word problem.

- I can set up appropriate equations for word problems.
- I can explain how the variables, numbers, and operations in the equations represent the word problems.
- I can justify my equations by checking the reasonableness of my solutions.

Midway through the lesson, Mr. Carlton pauses the class to further clarify the success criteria and points back to the problems they've been working on, such as the one in the following box:

On e-Music, you can download an entire album for $12, and individual songs are $1.25 each. Jake has an e-Music gift card with $25 on it. If he buys one album, how many additional individual songs can he buy?

1. What should the variable represent in this problem?

2. Write an algebraic equation that could be used to help you solve this problem.

Mr. Carlton observed that his students are least clear about the second success criterion, so he focuses there. First, he gives students some time to consolidate their learning during the lesson.

Mr. C: *So let's take a look at the second success criterion here. Turn to your partner, and talk for 30 seconds about how you would explain how the variables, numbers, and operations in the equations represent the word problems. Then write down what you think would be a strong response for this success criterion. Put your names on the back of your paper, please.*

Using their responses, he takes stock of what the class understands. After the students have talked and written their responses, he collects the papers and randomly selects three to read. This time, he chooses a few papers randomly; other times, he chooses papers of specific students who he considers representative samples of the rest of the class. He sees that students can explain how the variable is represented in the word problem but are not solid on explaining the operations.

From the information gathered, he gives them whole-class feedback:

Mr. C: *I've only looked at the work from three pairs so far, and in those three, I'm seeing that all of them have strong explanations of how the variables relate to the word problem. (He lines them up under the document camera so students can see all three responses, and he points out what it is about them that meets the success criterion.) I'm also noticing that the explanation of how the operations relate to the word problem seems to be where you're struggling a bit.*

Finally, he uses the opportunity to clarify the learning intention and the second success criterion:

Mr. C: *Can anybody share one way that they think they've been explaining how the operations in the equations represent the word problem?*

Ja'laine: *You have to multiply, and you have to add.*

Mr. C:	*So you're saying you need two different operations. How do you know that?*
Ja'laine:	*Because you need to multiply $1.25 times the songs to find out what he spends on songs. Then you add it to the $12 to get $25 total.*
Luis:	*And you have to subtract.*
Mr. C:	*Why do you have to subtract?*
Luis:	*Because you subtract $12 from $25, and that's how much money he can use to buy songs.*
Mr. C:	*(Seeing an opportunity to use each student's comment to help clarify the success criterion.) So, Ja'laine did a nice job explaining how the multiplication and addition were needed to set up the equation. That's what we're looking for in this success criterion—think about how the operations fit in the equation that you're creating to represent the problem. Luis went the next step beyond and correctly told us that you could subtract as part of solving the equation, but that's part of taking the next step to solve it and isn't part of this particular success criterion. So, Luis, you actually went beyond this success criterion.*

The conversation continues, as Mr. Carlton clarifies that students' explanations need to make clear what the numbers represent in the word problem as well. With this additional clarification, he then encourages them to keep working.

In this example, Mr. Carlton paused to revisit the success criterion intending two benefits—to give his students a moment to consolidate their learning and to take stock of where they are. The results allowed him include the other benefits for revisiting the success criteria during the lesson, but it is not necessary to try to meet all four at any one time; you might fulfill only one, or sometimes more—but not all—of them. The important idea here is that in revisiting the success criteria, you are giving students opportunities to strengthen their learning.

There are various strategies we encouraged teachers to use as they started learning to revisit the learning intention and success criteria with students during a lesson. The three Revisiting Strategies presented in the Framework resources on the companion resource website each provide suggestions for ways to build in opportunities in the midst of your lesson to enact the benefits for revisiting:

- *Taking Stock to Clarify and Consolidate:* This strategy describes one way to structure an opportunity to take stock of students' understanding in the middle of your lesson.
- *Gathering Evidence to Learn More:* This strategy describes one way to pause to gather more evidence in the midst of your lesson, with references to a number of different evidence-gathering techniques that can be found in the book *Mathematics Formative Assessment: 75 Practical Strategies for Linking Assessment, Instruction, and Learning* by Keeley and Tobey (2011).
- *Providing Formative Feedback to Guide and Move Forward:* This strategy describes a structure for providing whole-class formative feedback in the middle of your lesson, with references to a number of different feedback-related techniques that can be found in the book *Mathematics*

Formative Assessment: 75 Practical Strategies for Linking Assessment, Instruction, and Learning by Keeley and Tobey (2011).

Note that any of the Sharing Strategies provided earlier can also be adapted for use as a Revisiting Strategy as well.

 Classroom Resources: Go to Resources.Corwin.com/ CreightonMathFormativeAssessment to access these resources.

 Framework for Using LI and SC With Students. This printable file describes a way to help students understand and use learning intentions (LI) and success criteria (SC).

 Revisiting Strategies. This printable file describes strategies for revisiting learning intentions and success criteria in the midst of a lesson.

 Teacher Summary Cards for the Strategies. This printable file provides large-sized cards with step-by-step summaries of teacher moves for the Revisiting Strategies.

 Classroom Material: Go to Resources.Corwin.com/ CreightonMathFormativeAssessment to access this resource.

 Student Summary Cards. This printable file provides large-sized cards with step-by-step summaries of student moves for the Revisiting Strategies. This can be distributed to each student or pairs of students for use when learning about a strategy.

So far, we've illustrated revisiting the learning intention and success criteria in the midst of the lesson; revisiting the learning intention and success criteria at the conclusion of your lesson can help bring some closure to the lesson. This concluding revisit is particularly valuable to help students pause for a moment to synthesize their thinking from the lesson and to articulate aloud or in writing what they are taking away from the lesson. It also doubles as valuable evidence for you to review when considering where to focus when you next meet with the class. The following Wrapping-Up Strategies for revisiting the learning intention and success criteria at the conclusion of a lesson are provided on the companion resource website, though some serve other purposes as well, as we discuss a little later in the chapter.

- *Using Exit Tickets:* This strategy describes a way to create an exit ticket that focuses on your learning intention. (Note: The Using Exit Tickets resource from Chapter 3, which focuses on analyzing completed exit tickets, provides a good way to follow up with the evidence you gather with this strategy.)
- *X Marks the Spot:* This strategy also helps students practice self-assessment in relation to the success criteria.

- *Reflect Aloud:* This strategy also helps you model for students what it looks like to share their thinking.

 Classroom Resources: Go to Resources.Corwin.com/ CreightonMathFormativeAssessment to access these resources.

 Framework for Using LI and SC With Students. This printable file describes a way to help students understand and use learning intentions (LI) and success criteria (SC).

 Wrapping-Up Strategies. This printable file describes strategies for revisiting learning intentions and success criteria to help close or wrap up the lesson.

 Teacher Summary Cards for the Strategies. This printable file provides large-sized cards with step-by-step summaries of teacher moves for the Wrapping-Up Strategies.

 Classroom Material: Go to Resources.Corwin.com/ CreightonMathFormativeAssessment to access this resource.

 Student Summary Cards for the Strategies. This printable file provides large-sized cards with step-by-step summaries of student moves for X Marks the Spot in the Wrapping-Up Strategies. This can be distributed to each student or pairs of students for use when learning about a strategy.

 Reference Resource: Go to Resources.Corwin.com/ CreightonMathFormativeAssessment to access this resource.

 Using Exit Tickets as Evidence. This printable file describes a process, when examining exit tickets, for focusing on trends in student responses rather than on individual responses.

A Note on a Common Pitfall: But I Just Don't Have Time to Revisit!

Given the pressures of getting through the material you're responsible for teaching, it's easy to skip all this revisiting of learning intentions and success criteria because you feel that you don't have enough class time to devote to doing it. However, there's one very good reason to consider giving it time: *revisiting actually strengthens your students' mathematical learning.* As you revisit the learning intentions and success criteria, you're creating opportunities for your students to pause and consolidate their learning. We can't emphasize enough the importance to students—particularly to struggling learners but important to all learners—of pausing to consolidate their learning. It doesn't need to take much of your class time, but making room for it on a regular basis can result in using the time you *do* have remaining more productively. Any of the revisiting strategies or wrapping-up strategies provided here can help you provide this opportunity for your students.

As you consider whether to take the time and if so, how to take the time, the good news is that there are a variety of ways to do this, embodied in some of the strategies mentioned earlier, that don't require a lot of time. The better news is that as you do this over time, you are likely to see students engaging more productively during the class time that you *do* have.

TEACHERS' VOICES

It has become clear recently that the better I get at sharing and revisiting the learning intention and success criteria, the more students take ownership of their own learning. It seems the more I share the learning intention and success criteria, the better I get at writing them and the easier it is for the students to access them throughout the lesson as a real learning tool. When I miss the mark, the students themselves are sometimes able to (both directly and indirectly) help me to reword or reframe the success criteria to match what is really going on in class. Also, when class focus needs to change because of student gaps or misunderstandings, I am in a better position to quickly come up with success criteria that apply to the shift. And I am better equipped to know quickly in my mind what it would look like if the students were actually meeting the success criteria.

Helping Students Learn About Evidence

Earlier in the chapter, we presented two things that students need to learn, related to the critical aspect of Eliciting and Interpreting Evidence:

- To talk about their mathematical reasoning, not just their answers, so that you, and they, can accurately uncover how they are making sense of the mathematics
- To use the success criteria to self-assess their work to know in what ways they've met the success criteria and to what extent

The following recommendations and resources can help you work with your students to develop both of these skills and habits.

> **Recommendation 3: Model for students what it looks like to share their thinking, and give them opportunities to practice doing so.**

In Chapter 3, we talked about two important elements of the students' role in relation to evidence: participation and self-monitoring. This recommendation focuses on the participation part of their role: To participate in your efforts to gather evidence of their thinking, students need to be able to share the kind of information you're trying to gather from them about *how* they are making sense of the mathematics content, not simply *whether* they have learned it.

There are a variety of strategies you can model to help your students learn to share their thinking. These strategies include

- doing a think-aloud in which you provide a monologue of your thinking process as you solve a problem or try to create a written explanation of a math idea;
- comparing strong versus weak student responses from either fictitious students or from your actual students' work; and
- providing sentence starters or writing prompts for students to complete.

You might use these individually or combine them. For example, you might do a think-aloud about completing a sentence starter that you want students to use. You can also build on the Reflect Aloud Wrapping-Up Strategy presented earlier in this chapter to model a particular type of think-aloud in which you take on the role of a student and take stock of where your learning is at a given moment. In this strategy, you reflect aloud on the tasks you've done so far in the lesson, how those tasks relate to the success criteria, and how you think you (the student) are doing both in terms of your mastery and your confidence. In this way, you are modeling the internal self-assessment dialogue that you want your students to learn to have for themselves.

TEACHERS' VOICES

I tried the Reflect Aloud strategy with a learning intention that was about students learning to compare fractions in various ways. . . . As I reflected on my own use of the methods mentioned in the success criteria, I said that I seemed to be relying on certain methods more than others and wondered why that was. I also included a few example problems as evidence for each criterion. When I asked the students how their own self-assessment compared with mine, many of them agreed. We discussed this and realized that finding a common denominator (which was one of the criteria) was the toughest method to learn for comparing fractions. Many students reported then that they needed a little more practice with that criterion. I asked students to show me a thumb up, side, or down for that particular criterion, and the next day I met with the side and down students in a small group.

The key to all of these strategies is that there is built-in scaffolding for the students to help guide them in the right direction. In providing a structure for them to mimic, you may be able to help them eventually internalize that structure that they can use for themselves.

Recommendation 4: Help students learn to self-assess using the success criteria.

You can help students learn to assess their progress against the success criteria by modeling what self-assessment looks like. With any strategy or tool you choose to use to model self-assessment for your students, it's important to help your students learn to provide evidence of their self-assessment. As we've mentioned, many students initially will simply rate themselves based on their confidence level or even their reaction to whether they liked a lesson or activity or not. Helping them learn to point to evidence of their self-assessment can begin as simply as encouraging students to complete the phrase, "because _____" when they provide a self-assessment. For example, having to articulate "I know I met the success criteria fully because ____" or "I partially met the success criteria because ____" causes students to think a bit more about what it means to meet the success criteria, instead of thinking simply in terms of "I got it/I didn't get it."

TEACHERS' VOICES

I used traffic light dots recently before a test. Students completed a review sheet, and I had them color code it red (meaning I'm feeling lost), yellow (meaning I sort of understand it but need a little more help or practice), or green (meaning I'm solid on this idea) by drawing a dot next to each problem. Then I sorted through the papers and identified which problems students had marked as yellow and red. As a class the next day, we went over the concepts and problems that many students had marked yellow. Then I met in small groups with students who had marked something red to review those concepts. I think it was beneficial for students to see how I used their self-assessments to inform instruction and how it affected the activities we did in class. Seeing that I valued their honesty as well should encourage more students to participate and stay involved more in the future as well.

Another way you can model using the success criteria to assess work is to provide examples of student work (real or fictitious) and discuss them with students to determine what makes an example high quality in terms of the success criteria. As students and teacher discuss what makes one sample a high-quality sample and what makes another low quality, the students develop a common understanding of what is considered high-quality work and can then apply those standards to their own work.

TEACHERS' VOICES

I have a wall for "Wow work" and put samples related to class goals. The Wow Wall became a place where students chose to put their work. I still placed samples there and used them as links to goals, but students could choose work themselves and often did so. Next year, I should add the associated learning intention and success criteria as well to the wall.

In Chapter 3, we talked about the potential confusion students can have about what evidence means—whether it refers to their answers, their actual work on a problem, how well they feel they've learned something (their confidence), or even their reaction to whether they liked a lesson or activity or not. They need help learning to understand the difference between sharing their mathematical reasoning and results and indicating to you how confident they feel about their learning. One way you can help them learn to do this is to be explicit about this difference and to practice self-assessing both their mastery and their confidence separately.

Some of the resources we've already mentioned can also be used to model for students what it means to self-assess their work and their thinking. For example,

- the *Taking Stock to Clarify and Consolidate* Revisiting Strategy describes a way to model for students one way to self-assess. It can be usefully paired with the *Reflect Aloud* Wrapping-Up Strategy.
- the *X Marks the Spot* Wrapping-Up Strategy and accompanying template provide a tool students can use to assess themselves in relation to the success criteria, with attention given to the evidence for the self-assessment. Variations on the strategy are suggested as well.

In addition, the Self-Assessment Poster, available on the companion resource website, can be used to provide students with a set of prompts they can choose from to describe how well they understand the current mathematics content. Each prompt ends with "Here's why," to encourage students to use the evidence of their work to support their choice of prompt.

 Classroom Material: Go to Resources.Corwin.com/ CreightonMathFormativeAssessment to access this resource.

 Self-Assessment Poster. This large-format PDF provides self-assessment prompts to post in your classroom. The file can be printed poster size through your local office supply or copy center.

Helping Students Learn About Formative Feedback

Earlier in the chapter, you saw that formative feedback is involved in three of the four parts of self-regulation: Where am I currently in relation to those goals? What do I need to do next to be able to meet them? and taking action to move learning forward. Therefore, helping your students learn the following things related to formative feedback is an important foundation for self-regulation:

- What formative feedback is and how to use it to support the self-assessment process
- To use peers as well as their teacher as resources to help them when they are stuck
- To look for the part of the feedback they receive that has hints, cues, or suggestions for productive next steps

- To use the hint or cue provided in the feedback to take some next steps in their work

The next recommendation and related resources can help your students learn about these things.

> **Recommendation 5: Explicitly talk to students about what formative feedback is and how to use it.**

As with learning intentions and success criteria, one way to help students gain a better understanding of formative feedback is to devote some time to explicitly teach students what it is. All students are accustomed to getting feedback from their teachers, so the new learning for students here is specifically how formative feedback is directly related to the success criteria and how it therefore can help them understand how they are doing in relation to those success criteria. As we mentioned in Chapter 4, we worked with teachers to develop a lesson plan that introduces students to formative feedback. You can post the Feedback Poster used by the lesson on a bulletin board or wall, so in future lessons you can remind students to pay attention to all parts of formative feedback.

 Classroom Resource: Go to Resources.Corwin.com/ CreightonMathFormativeAssessment to access this resource.

 Introduction to Formative Feedback (for Students). This printable file provides a lesson plan to introduce your students to formative feedback.

 Classroom Material: Go to Resources.Corwin.com/ CreightonMathFormativeAssessment to access this resource.

 Feedback Poster. This large-format PDF can be posted in your classroom for ongoing reference to the characteristics of feedback. Suggested for use with the "Introduction to Formative Feedback (for Students)" lesson. The file can be printed poster size through your local office supply or copy center.

Once students understand what formative feedback is, you can help them learn to use feedback they receive to help them identify how they are doing in relation to the learning intention. First, you can model the process of using success criteria as the basis for providing formative feedback. Using a template or structure that becomes familiar to students over time, such as the resource listed later, you can model ways to provide feedback to students, then gradually give them opportunities to provide feedback to others using that same structure. The sentence starters provided in the following resource point to the success criteria to help the receiver of the feedback understand three things:

1. Are there success criteria I have met? If so, how?

2. Are there success criteria I have not met? If so, why?

3. For any that I have not met, what suggestions do you have for what I might do or try next?

Classroom Material: Go to Resources.Corwin.com/ CreightonMathFormativeAssessment to access this resource.

 Feedback Templates. This printable file includes templates that you can use to provide feedback to your students and that they can use to provide peer feedback.

You can also provide opportunities for students to practice providing formative feedback to a fictitious other student and then to peers. Some teachers may have had unproductive experiences with peer feedback and may not feel it's worth the time or effort. However, if that has been your experience, we urge you to give it another chance. Students need to learn how to give a hint or clue that's useful, aiming for a balance between not too much or not too little. Striking this balance may seem like a tricky thing to do, but students can learn relatively easily to at least avoid doing the work for their peer in the feedback and to avoid simply telling them whether or not they're correct. As they learn to use the success criteria as the "stuff to talk about" in their feedback, they quickly learn to get more specific in a useful way.

Providing formative feedback to peers can provide a really important practice run for students as they learn to do it for themselves. As students practice giving feedback to peers, they initially focus on the question, "How is this other student doing in relation to the success criteria?" You can then help them learn to look at their own work and ask "How am *I* doing in relation to the success criteria?" and to be able to answer that question using evidence compared to the success criteria.

As with introducing students to formative feedback, devoting some time to explicitly teaching students how to provide feedback to peers helps the experience to be productive for all involved. Students may have been asked in the past to provide some kind of feedback to peers, but without guidance on what makes good feedback, they may have tended to focus on remarks about the neatness or clarity of the work or on whether the student liked it, rather than on important mathematical elements of the learning goal. So the new learning here for students is about using the success criteria to provide feedback that addresses the three questions mentioned earlier.

We have provided a second follow-up lesson, also referenced in Chapter 4, which addresses learning to give formative feedback to a peer. You can also use many of the previous formative feedback classroom resources we've mentioned to help students learn to pay attention to the hint, cue, or model.

**Classroom Resource: Go to Resources.Corwin.com/
CreightonMathFormativeAssessment to access this resource.**

Peer Feedback. This lesson introduces providing feedback to peers.

As students gain experience providing feedback to peers, some may begin seeking each other out on their own when they are getting stuck, but others will need encouragement. You may also have a number of structures that you can use to have your students work with peers or seek out peers for help when they are stuck. If so, you can adapt those structures to have students focus on the success criteria as a resource to guide their work. In addition, many of the resources we've already listed also serve as a possible structure for using a peer as a resource. In particular, the Feedback Poster and Feedback Templates help students point back to the success criteria as a basis for the feedback they might give to their peers.

Helping Students Learn About Taking Action

In answering the third self-regulation question, "If I have not yet met the goal, what do I need to do next to be able to meet it?" and then completing the final step, taking action, there are two more things that students need to learn:

- To use various resources (such as their notes, different problem solving strategies, textbooks, articles, and tutorials on the Internet) to move their learning forward
- To use the hint, cue, or model provided in the feedback to take some next steps in their work

The first of these is noteworthy as seeming to be outside the formative assessment process—but in a self-regulating student, this action can be thought of as *internal feedback*. This is a valuable skill for lifelong learning, useful even as students leave school and no longer have an official teacher to help them. As with much of self-regulation, helping students learn to use the resources at their disposal takes time, but the rewards for you, for them, and for their future teachers are immeasurable.

But all students need external feedback too, from time to time. As students become more familiar with what formative feedback is, and the importance of paying attention to the hint, cue, or model that provides a suggestion for next steps, you can help students learn to follow through with taking action on any feedback they do receive. This last part of the self-regulation process is where the real impact of receiving feedback lies for students; without it, the feedback becomes merely information to file away that may or may not have any value in moving students' learning forward.

We provide here one final recommendation and resources to help you teach your students how to take action.

> **Recommendation 6: Teach students structures for taking action, and use them frequently enough so that they become familiar to students.**

Earlier in this chapter, we asked you to think about the kinds of hints, cues, and models you might give to students as you provide formative feedback to them. To help students provide internal feedback, you might give them a list of things to try and then support them in using it. Figure 5.7 illustrates how one teacher decided to provide such a list, on the "When I Am Stuck" side of her Moving Learning Forward bulletin board:

Figure 5.7   Moving Learning Forward Bulletin Board

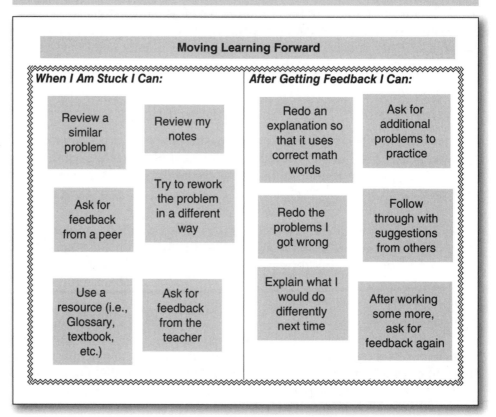

As you give feedback, you can turn to the list and say, for example, "I think you might have something in your notes that can help you. Look through your notes from yesterday, and see if you can find a similar problem that could give you some insight into this one." Over time, you can put this decision more and more into their hands: "Look at our list of things to try. What do you think might be helpful this time?"

Notice that the Moving Learning Forward bulletin board also has a side for what to do with feedback. Students may be unaccustomed to doing anything with the feedback they receive. Many students, particularly struggling students, are reluctant to take any initial steps to solving a problem—much less act on feedback—without the teacher's guidance, step by step through the problem. Some students may feel very helpless about knowing what to do next to improve their work or revise their thinking or about having the confidence that they can take action, even if they *do* know what to do. Learning about what formative feedback is, and helping students focus on the hint, cue, or model provided by that feedback, is one way we've already discussed that should help them past this feeling of helplessness. Encouraging actions such as those given on the bulletin board is another way to help them use their feedback.

However, learning to work with feedback will be a change of habit for many students. To make an effort to make this change, we recommend giving your students a structure to use to guide their steps, in the form of a brief instructional routine that you can use any time you want students to do something with the feedback they've received. We've provided one example of such a structure for this purpose on the companion resource website. You may find you want to adapt it to better fit your teaching style or your students' needs, or you may develop a different structure altogether.

Classroom Resource: Go to Resources.Corwin.com/ CreightonMathFormativeAssessment to access this resource.

Feedback Response Strategy. This printable file describes a strategy that you can give your students to help them respond to feedback.

The key to success with structures such as this, or the Moving Learning Forward bulletin board, is using them often enough that they become routine for students. The important thing here is to create something for your students to use to help them learn to act, on the feedback that they receive or that they give themselves, and use it regularly so that they can begin to internalize the steps and use them independent of the prompts in the routine.

Planning Resource: Go to Resources.Corwin.com/ CreightonMathFormativeAssessment to access this resource.

Formative Assessment Planner Templates. This printable file includes two versions of templates to help you create a formative assessment plan for a lesson. One version includes a portion in which you can plan to support students in self-assessment and self-regulation.

SELF-REGULATION: IT'S NOT A LINEAR PROCESS

All the recommendations and resources we've described so far are intended to help you support your students in becoming more self-regulating with their mathematics learning. We've tried to break the process students go through down into recognizable parts and discuss them somewhat discretely. We've described the steps in terms of four parts represented in the diagram, and we have moved through them in an orderly left-to-right motion. In actuality, students' movement through the self-regulation process is not at all linear but instead moves back and forth between different parts. For example, as students gain greater clarity about the success criteria and the learning intention, it can cause them to reevaluate their success in meeting them, as well as what they now need to do to actually reach the learning intention. As they get feedback, they are better able to refine their work and thinking.

Let's look at a vignette that illustrates what this might look like. As you read the vignette, pay attention to how students are moving to different parts of the self-regulation process, not necessarily in a linear way but in a way that weaves the parts of the process together.

At the start of a 2-day lesson on solving simple equations, Mr. Washington introduces the learning intention (LI) and success criteria (SC) to his students, using the Clarification Strategy to focus on clarifying the meaning of equivalence and equation.

LI: It's important to maintain equivalence on both sides of an equation when you are solving it.

SC 1: I can solve an equation using steps that maintain equivalence.

SC 2: I can describe what operations will maintain equivalence when solving an equation.

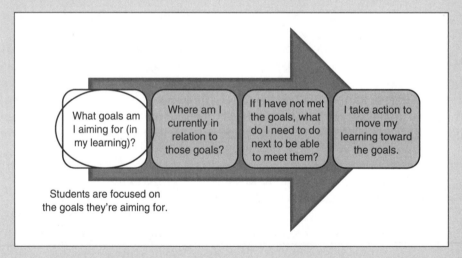

Students are focused on the goals they're aiming for.

In the first part of the lesson, students learn about a balance-scale analogy to help them think about maintaining equivalence as the quantities on both sides of the balance scale change. After the activity, Mr. Washington leads a brief discussion about the question, "When you're solving an equation, why do you need to do the same thing to both sides?" When the student responses don't sound very promising, he writes the following problem on the board and asks students to write on their individual whiteboards "yes" or "no."

$$3x + 4 = 5x - 6$$
$$-\ 4 \qquad +\ 6$$

Do these steps maintain equivalence on both sides of the equation?

A quick scan of the students' responses tells him that most of the class is incorrect. Seeing more incorrect yesses than correct nos, he says, "Most of you think that subtracting 4 and adding 6 will maintain equivalence, but the correct answer here is no. So we're not yet meeting our success criteria; let's figure out why."

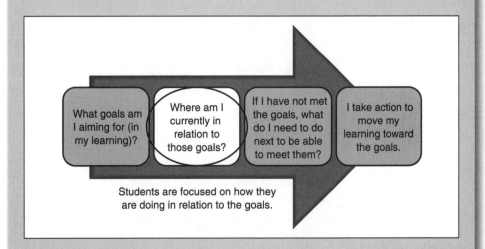

He points to the success criteria and asks, "What does it mean to maintain equivalence?" After a brief discussion, it becomes clear that some students thought it meant simply doing the inverse operation to cancel out the constant term. As a result, Mr. Washington is able to clarify the meaning of equivalence and to talk further about when and why you use the inverse operation.

Mr. Washington gives students some additional problems to try for homework and also asks them to write a sentence in their own words about what it means to maintain equivalence.

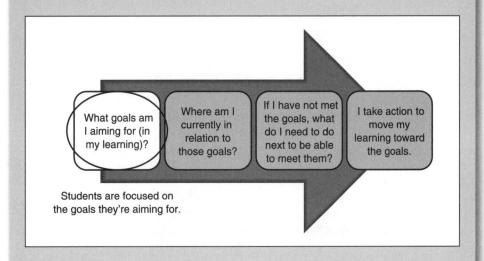

At the start of the next class, Mr. Washington has students do a think, pair, share in which they share their sentences with a partner, then share out in a brief discussion their current understanding of the phrase maintain equivalence.

(Continued)

(Continued)

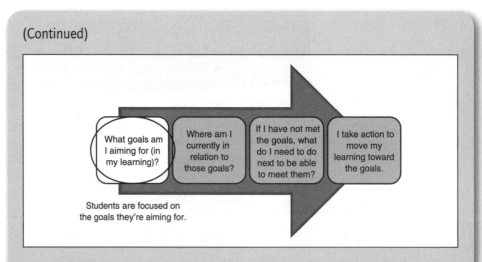

Students are focused on
the goals they're aiming for.

Mr. Washington sees that more than half of the class is now thinking correctly, but some students still are confused about what steps to do and why. He offers three possible choices of activities for this lesson—working on more difficult problems, just practicing some more with the current kind of problems, or meeting in a small group with him for some help—and asks students to choose which activity they want to spend their time on this class period.

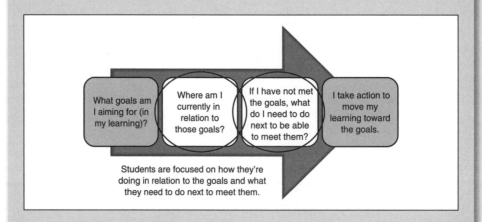

Students are focused on how they're
doing in relation to the goals and what
they need to do next to meet them.

In the final 5 minutes of class, Mr. Washington briefly returns to the learning intention to talk about how the activities of the past 2 days relate to it, and he gives students an exit ticket that says "When you solve an equation, what can you do that will maintain equivalence in the equation? Explain how your answer maintains equivalence."

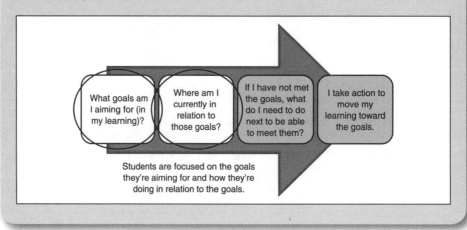

Students are focused on the goals
they're aiming for and how they're
doing in relation to the goals.

In this vignette, Mr. Washington provides opportunities for his students to reflect on their learning in relation to the learning intention and success criteria, and those reflection opportunities—coupled with feedback from him—help his students gain a clearer understanding of the concept of equivalence in equations. This understanding goes beyond simply understanding how to write out the correct series of steps to solve an equation; it also encompasses understanding why certain steps are performed the way they are, so they are no longer just a rote procedure to follow. And the more that students understand the *whys* underlying the *how tos*, the more likely they are to retain the information and be able to apply it in a variety of situations.

CONCLUSION ■

In this chapter, we've discussed how a more self-regulating student thinks and behaves in comparison to a less self-regulating student, painting a picture of the eventual behavior and internal dialogue that we hope students will build as a result of experiencing formative assessment practices in math class. Moving toward this goal of increasing your students' ability to self-regulate involves much new learning for students along with planning and support on your part. We've provided a collection of resources and tools in this chapter to support your efforts to help your students move in this direction, and links to other resources that have been created by others can be found on the website under our Resources section.

Pulling this all together may feel like a big task, and while that may be true, it is also a manageable task with terrific payoff in the form of students who are more engaged in mathematics class and who feel more success and more confidence in their ability to learn mathematics.

TEACHERS' VOICES

One of my students began the year as a conscientious participant. He always did his homework, and he often participated in whole class discussions. However, his calculations were often inaccurate, and he had many embedded misconceptions about math.

As we progressed into the year, his confidence dropped to a very low level. His mathematics journal, where I have students reflect on their class often, is full of "I did not understand the directions" and "It was slightly difficult" and "I need help." His mother called at one point to tell me there was great frustration at home over homework and math in general.

Later in the year, these comments appeared in his journal: "I was successful because I understood the problem." "I work better in a group because we share ideas." "I need to slow down as I work." And as we worked on sample items from the state test, he wrote, "It bothers me if the wording's weird and I can't ask for clarification."

At the end of the year, I had my students write in their journal to the incoming sixth graders. Based on the way he described math, I would say that formative assessment practices have really made a difference for this student. He is not only taking ownership of his learning, he has learned to self-advocate in class and he has found strategies for getting through the rough spots of math—when understandings are new and unclear—to when the understandings are solid and well developed.

In our final chapter, "Moving Toward Implementation," we talk about how to get yourself started and how to maintain your efforts at weaving all the aspects together. However, before we get there, we need to address our two important supporting aspects that provide the necessary backdrop to implementing formative assessment practices successfully: learning progressions and classroom environment. We discuss those supporting aspects in the next two chapters.

Recommendations in This Chapter

1. Explicitly talk to students about what the learning intention and success criteria are and how you are using them.

2. Use a variety of ways to clarify for students the meaning of the learning intention and success criteria, including near the start of the lesson and during the lesson.

   a. Draw students' attention to the learning intention and success criteria initially to clarify any language and discuss any potentially confusing points.

   b. Draw students' attention back to the learning intention and success criteria at numerous points throughout the lesson both to help them stay focused on the intended learning for the lesson and to help further clarify the intended learning.

3. Model for students what it looks like to share their thinking, and give them opportunities to practice doing so.

4. Help students learn to self-assess using the success criteria.

5. Explicitly talk to students about what formative feedback is and how to use it.

6. Teach students structures for taking action, and use them frequently enough so that they become familiar to students.

Reference Resources: Go to Resources.Corwin.com/
CreightonMathFormativeAssessment to access these resources.

 Chapter 5 Summary Card. This printable file provides a Summary Card of the recommendations in this chapter.

 Formative Assessment Recommendations. This printable file provides a summary of all the recommendations in this book.

RESOURCES ■

The following sections present some resources that can help you implement the recommendations. Each of these resources is referenced in earlier sections of the chapter, but here we provide a consolidated list. All resources can be found at Resources.Corwin.com/CreightonMath FormativeAssessment.

Learning Resources

In addition to the resources mentioned in this chapter, videos are available to support your learning about formative assessment practices related to developing student ownership and involvement.

Reference Resources

These resources summarize key ideas about student ownership and involvement:

- **Images of Posted Learning Intentions and Success Criteria** includes images of the various ways teachers have posted learning intentions and success criteria for reference and use during a lesson.
- **Chapter 5 Summary Card** is an index-size Summary Card that includes all the recommendations described in this chapter.
- **Formative Assessment Recommendations** is a printable PDF that includes all the recommendations from this book.

Planning Resources

This planning tool supports your lesson planning when integrating formative assessment into your mathematics lessons:

- **Formative Assessment Planner Templates** can be used for creating a formative assessment plan for a lesson. One version includes a portion for you to consider how you will support students in self-assessment and self-regulation.

Classroom Resources

These resources illustrate various classroom strategies to develop student ownership and involvement.

- **Introducing Students to Learning Intentions and Success Criteria** is an introductory lesson to help students understand the purpose of learning intentions and success criteria.
- **Framework for Using LI and SC With Students** describes a way to help students understand and use learning intentions (LI) and success criteria (SC) during a lesson. It describes actions for both teacher and students to take at the start of the lesson, at midway points during the lesson, and at the end of the lesson.

- **Sharing Strategies** describes several strategies for use when introducing students to LI and SC for the first time.
- **Revisiting Strategies** describes several strategies for use when revisiting LI and SC throughout the lesson.
- **Wrapping-Up Strategies** describes several strategies for use referring to LI and SC to wrap up a lesson.
- **Teacher Summary Cards for the Strategies** is a set of large-size cards that provides a step-by-step summary of *teacher* moves for the strategies described earlier.
- **Self-Assessment Techniques** lists formative assessment classroom techniques (FACTS) related to student self-assessment that can be found in the book *Mathematics Formative Assessment: 75 Practical Strategies for Linking Assessment, Instruction, and Learning* by Keeley and Tobey (2011).
- **Introduction to Formative Feedback (for Students)** is a lesson plan for introducing the characteristics of formative feedback to students.
- **Peer Feedback** is a lesson plan for students about introducing and practicing how to give formative feedback to peers.
- **Feedback Response Strategy** describes a strategy that you can give your students to help them respond to feedback.

Classroom Materials

These resources can be copied or re-created to be used in your classroom.

- **Self-Regulation Poster** can be posted in your classroom to encourage students to think like a self-regulating learner. The file can be printed poster size though your local office supply or copy center.
- **Student Summary Cards for the Strategies** is a set of large-sized cards that provides a step-by-step summary of *student* moves for the strategies described earlier. This can be distributed to each student or pairs of students for use when learning about a strategy.
- **Self-Assessment Poster** can be posted in your classroom to provide prompts that encourage students to provide evidence from their work in their self-assessment. The file can be printed poster size though your local office supply or copy center.
- **Feedback Templates** for using the feedback techniques described earlier:

  o Providing Feedback includes Feedback Sandwich and Peer to Peer Focused Feedback
  o Reflecting on Your Use of Feedback includes Feedback to Feed-Forward and Planning from Feedback

- **Feedback Poster** can be posted in your classroom for ongoing reference after introducing students to the characteristics of feedback through the "Introduction to Formative Feedback (for Students)" lesson. The file can be printed poster size though your local office supply or copy center.

6

Using Mathematics Learning Progressions

We've now addressed the four critical aspects of implementing formative assessment: (1) learning intentions and success criteria, (2) eliciting and interpreting evidence, (3) formative feedback, and (4) student ownership and Involvement. In the next two chapters, we turn our attention to the two supporting aspects that provide a necessary foundation to these four: learning progressions and classroom environment. In this chapter, we look at what learning progressions are, why they're important, and how they support your implementation of formative assessment practices.

In Chapter 1, "Using Formative Assessment to Build Student Engagement in Mathematics Learning," we identified a learning progression as "an articulation of a pathway through which understanding of content evolves, from basic to more sophisticated understanding." This means learning progressions are about the development of mathematical content. You may be wondering why a discussion of mathematics content via learning progressions is addressed so late in the book, when attention to mathematics content is so fundamental to any mathematics instruction and most especially to effective use of formative assessment practices. The answer is this: Thinking about mathematics content is fundamental to any mathematics classroom, so many teachers already have given this a lot of thought. The particular question we want to take up is what is it about learning progressions that is particular to the implementation of formative assessment? To pose this question, we needed to first establish what implementing formative assessment means, which we hope we have done in the previous chapters. (Note that the same may be said of our other supporting aspect: classroom environment.)

So what *is* it about learning progressions that is particular to the implementation of formative assessment? We'll trace this relationship throughout the chapter using the example of Ms. Lucenta's seventh-grade mathematics classroom.

Ms. Lucenta has learned that the content standards for Grade 7 have changed for the coming school year and that she will now be responsible for teaching surface area and volume, a topic that she has not been responsible for in the past. She wonders how she will be able to incorporate this content, when she knows that her students have often been confused about how to find area and perimeter, and to keep track of which is which. They are supposed to have already addressed these topics in earlier grades, but students still come to her weak in these skills. If they aren't even strong on those skills, she wonders, how will they manage to master surface area and volume?

Ms. Lucenta's first thoughts turn to her students' ability to master skills related to finding area, perimeter, surface area, and volume, and her concerns are well founded. Having a solid understanding of two-dimensional linear and area measurement lays an important foundation for moving on to geometric measurement of three-dimensional figures. However, it's helpful to articulate not only the prerequisite skills that students need to master but also the prerequisite mathematics concepts. A learning progression provides a clear roadmap of how those concepts and skills relate to each other and develop for students.

■ WHAT IS A LEARNING PROGRESSION?

Much of the existing literature about formative assessment has underscored the importance of understanding a learning progression in order to effectively implement formative assessment practices, but there are a variety of differing definitions of what these progressions are. Some researchers have described them as "descriptions of the successively more sophisticated ways of thinking about an idea that follow one another as students learn" (Wilson & Bertenthal, 2005, p. 3), while others have described them as "a picture of the path students typically follow as they learn . . . a description of skills, understandings, and knowledge in the sequence in which they typically develop" (Masters & Forster, 1997, p. 1). The definition we use in this book focuses on progressions as descriptions of increasingly sophisticated levels of learning, and is most clearly articulated by Margaret Heritage, a leading researcher on formative assessment practices:

Learning progressions are generally considered to be descriptions of how students' learning of important concepts and skills in a domain develops from its most rudimentary state through increasingly sophisticated states over a period of schooling. While there can be differences among progressions in terms of their scope, breadth and level of granularity, most progressions lay out a pathway along which students will typically advance their learning in a subject area. This pathway is made up of important building blocks that represent a goal level of understanding or skill and articulates the changes in the level of understanding and skill that accrue with each successive step. (2011, p. 3)

A description of a learning progression may initially sound like it is synonymous with collections of state standards or curriculum maps. These items do have a connection, but they have important differences as well. The learning progression is a more general view of how learning evolves, which can be helpful in creating a curriculum map—which in turn is intended to help students meet the standards. Table 6.1 summarizes some important differences among these tools.

Table 6.1   Key Differences among Learning Progressions, Standards, and Curriculum Maps

State Standards	Curriculum Maps	Learning Progressions
Help answer the question: *What do students need to be able to do by the end of a particular grade?*	Help answer the question: *How well or in what way does our curriculum help us meet these standards or cover these topics?*	Help answer the questions: *What concepts and skills do students need to work with, and in what order, to learn about a particular topic?* *What ideas are foundational to the content I am teaching, and where will this content lead in future study?*
Describe learning outcomes. They lay out necessary concepts and skills to be learned by the end of a particular grade.	Map the location of and relationship between different concepts and skills in the intended curriculum.	Describe the learning that is typical as part of moving toward more sophisticated understanding of target concepts and skills.
Are grade-level specific—link specific standards to particular grade levels.	Are grade-level specific—connect target standards for a grade to elements of an intended curriculum.	Are not grade-level specific—do not link the timing of the learning of concepts and skills to particular grade levels.
Are used to indicate year-end learning goals for a particular grade level and to measure students' achievement at that grade level.	Are used to illustrate the relationship between a particular intended curriculum and a set of standards or topics.	Are used as a basis for articulating lesson-level learning goals, to make instructional decisions about appropriate next steps, and to more accurately diagnose gaps in student understanding.

Figure 6.1 The Evolving Formative Assessment Cycle (with learning progressions)

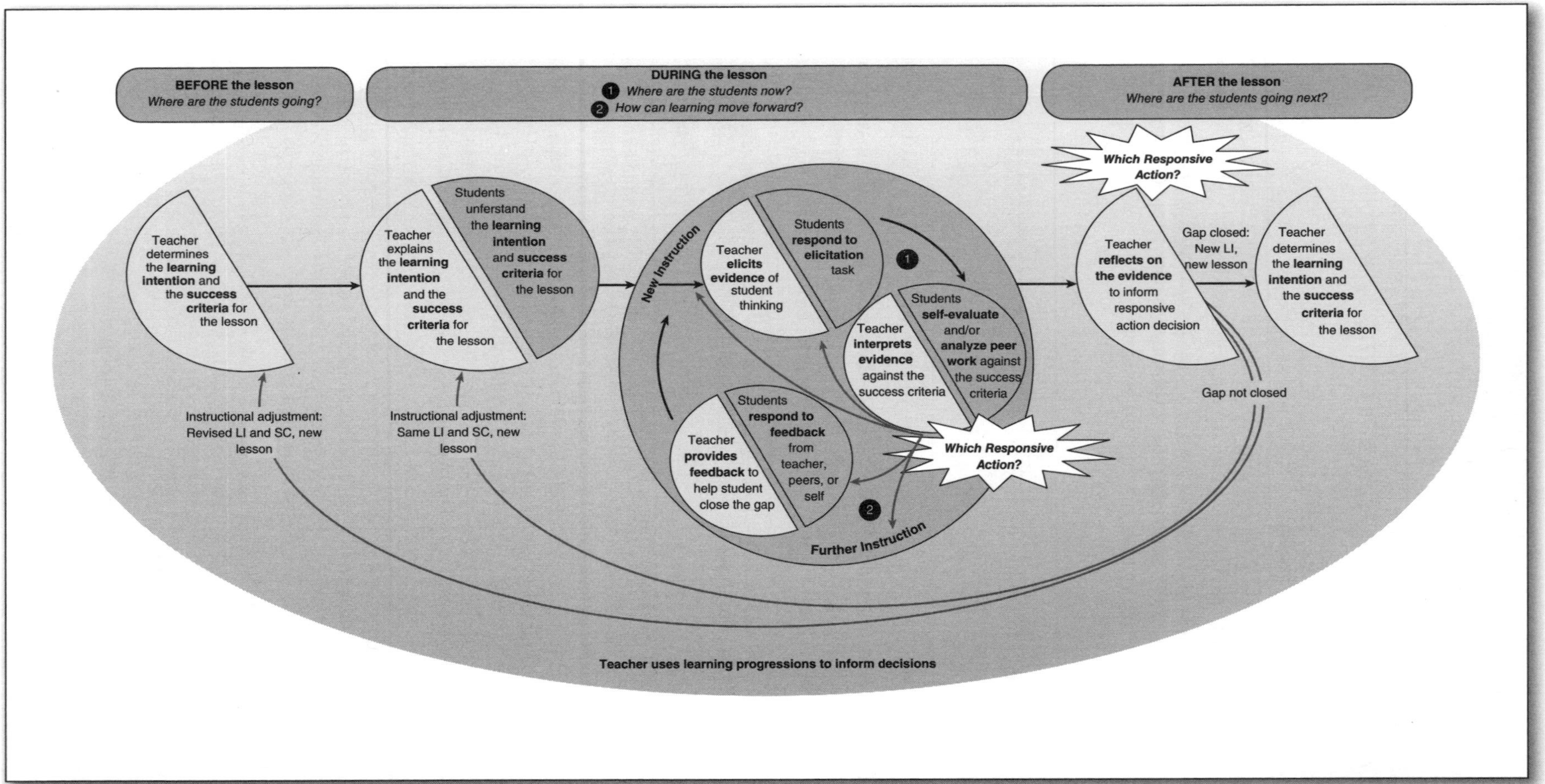

In our evolving Formative Assessment Cycle (Figure 6.1) we have added learning progressions in a "cradle" at the bottom, rather than within any of the half-circles where the critical aspects appear. This illustrates the supportive nature of learning progressions; as an articulation of mathematical content and how student understanding builds, they underlie all pieces of the formative assessment process.

Characteristics of Learning Progressions

Learning progressions can be described in many ways, from a list of text to diagrams of varying complexity. And judging from examples in the field, there does not appear to be any agreed-upon grain size for how general or how detailed a progression needs to be (though we say more later in this chapter about breaking down to a grain size that is useful for planning instruction).

Consider an example of a simplified learning progression for proportional reasoning. At the largest grain size, we could say that several important big ideas in proportional reasoning were

- constant differences,
- multiplicative patterns,
- multiplicative relationships, and
- unitizing and norming (for example, using a ratio as a mathematical unit to make comparisons).

If we were to try to articulate the important concepts to understand that were part of the big idea of multiplicative relationships, we could then list

- ratios,
- types of ratios (part-to-part, part-to-whole), and
- equivalent ratios.

If we were to peel back one more layer of our learning progression to uncover important understandings and skills that were part of understanding equivalent ratios, we might arrive at Figure 6.2.

Note that the concepts build from the bottom of the diagram to the top. Also, this diagram highlights equivalent ratios as part of the understanding of multiplicative relationships; each of the topics in our two lists above could be expanded in a similar way. We present the lists and diagram as one example, to give you some context for a number of specific characteristics of learning progressions.

- **A progression emphasizes concepts, and is supplemented by corresponding skills.**

   A progression can include both concepts (ideas) and skills (procedures, algorithms, strategies, etc.). However, if a mathematics learning progression is to describe a student's mathematics learning, then in the best of cases, that learning goes beyond the basic accurate execution of procedures and includes deeper conceptual

Figure 6.2 Learning Progression (based on the rational number research by Behr, Harel, Post, & Lesh, 1992; Cramer Post, 1993; Lamon, 1994; Lesh, Post, & Behr, 1988)

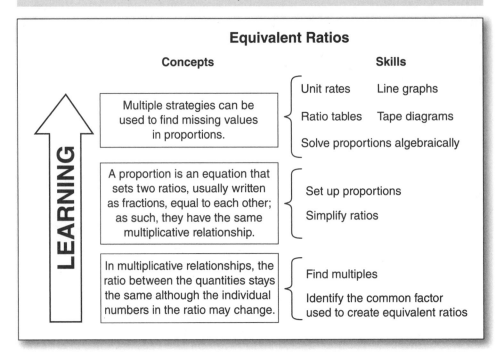

understanding. A useful learning progression will *emphasize* concepts, and then it will be supplemented by the corresponding skills that relate to those concepts. In the example, you can see the equivalent ratios box describes concepts on the left and the associated skills on the right.

- **A progression describes a gradual evolution of understanding from less sophisticated understanding to more sophisticated understanding.**

Note that *less sophisticated* and *more sophisticated* are relative terms. We use these rather than terms such as *basic* and *advanced*, which are dependent on the general age and experience of the students; for instance, something that is considered advanced understanding for a second grader may be considered basic understanding for a seventh grader. As you'll see, learning progressions are not tied to particular grade levels, so focusing on the progression from less sophisticated to more sophisticated helps avoid age and experience considerations.

Effective learning progressions are developed based on what is known in the field about students' learning of certain mathematics concepts. For example, the extensive work of researchers on the Rational Number Project (2002) over many years have determined quite a bit about how students make sense of rational numbers and have published numerous articles about students understanding in the area. Other researchers have defined elements of geometric thinking. This research into students' learning of various mathematics concepts describes ways that students make sense of the complex concepts in each of these areas of mathematics and therefore dictate something of an order to the various mathematics concepts in a progression. However, all teachers know that students do not learn the same things at the same time and in the same way. So any progression can only describe what you might expect students to do, knowing that any individual student may carve out a more unexpected path through the progression. For that reason, we prefer to describe learning progressions as *road maps*, in which there are definite highways that get you from Point A to Point B but that also provide a choice of route to get there. The route may sometimes involve passing through different towns, skipping a town, or backtracking. Different students will make sense of a collection of mathematical ideas in different orders and in different ways, but all can end up at the same destination.

- **Progressions do not assign particular concepts to particular grade levels; they are grade neutral.**

We mentioned above that students do not all learn the same things at the same time and in the same way. For example, if we step back to look at the development of middle school students' learning of the concept of proportionality across Grades 6 through 8, we know that some students will make sense of this concept in Grade 6, many others at different points across Grade 7, and others not until Grade 8. Learning progressions ignore grade-level assignments of particular topics because progressions differ in *purpose* from those documents that assign content to particular grade levels. Collections of content standards, scope and sequence documents, or curriculum maps that assign certain content standards to grade levels all define for the user what the target outcomes are for students at that grade level. A learning progression, on the other hand, describes *how* the understandings build to those outcomes (and beyond), describing key landmarks of understanding along the way. The particular grade level at which students develop certain understandings along the way are not relevant to the progression.

- **Progressions are not all the same in terms of their size, their breadth of concepts and skills, or in the level of detail described.**

A teacher working with early elementary students' understanding of place value may have need of a learning progression that describes children's conceptions of whole numbers and of addition and multiplication. A teacher working to build middle school students' understanding of rational numbers may have need of a broader, larger learning progression that describes concepts of whole number as well as rational numbers. However, a learning progression for the middle school teacher would become too unwieldy if it began at Square 1, describing early elementary students' emerging conceptions of whole numbers; the progression for the middle school teacher has to assume some prerequisite knowledge of whole numbers by all or most students and might begin somewhere in students' first experiences making sense of fractions and decimals from whole numbers. The important point here is that there is no prescribed starting and ending point of a learning progression; the scope and breadth of the progression have to serve the needs of the teacher using it for his or her population of students.

Progressions can also differ in the level of detail for different concepts. The elementary school teacher's learning progression might describe in greater detail the progression of learning of whole number concepts, while the middle school teacher's learning progression might touch more broadly on a few important prerequisite concepts in whole number that underlie fractions and decimals then move into more detail about concepts related to rational numbers.

■ HOW CAN YOU USE LEARNING PROGRESSIONS IN YOUR INSTRUCTION?

Learning progressions can serve three primary purposes for formative assessment practices in your instruction. We discuss each of these purposes in this section.

1. They provide mathematics background information to you, so that you can develop your own content knowledge as fully as possible. This knowledge informs every part of your mathematics instruction.

2. They help you plan instruction by mapping out important prerequisite concepts as well as prerequisite skills. A progression can inform your instructional decisions about where to go next both at the unit or chapter level as well as at the lesson level.

3. They help you diagnose potential student difficulties. Sometimes the mistakes that are apparent in a student's work can mask the real source of the error, if the source is conceptual. The information in learning progressions can suggest to you possible sources of students' difficulties so that you can more accurately pinpoint what students need.

Purpose 1: Using Learning Progressions to Provide Mathematics Background Information

Let's return to our vignette with Ms. Lucenta's seventh-grade classroom.

In preparation for having to teach surface area and volume, Ms. Lucenta decides to take a look at some existing learning progressions related to measurement and to geometry. She chooses to look at the Common Core State Standards learning progressions documents, available online. (She could also have chosen to look at any one of a number of relevant resources addressing development of mathematics content at the middle school level; see the "Books and Articles List" on the resource website for other suggestions.)

> ### COMMON CORE CONNECTION
>
> The Common Core State Standards have renewed attention to "learning progressions." The writing team started with progressions, "informed both by educational research and the structure of mathematics" (Common Core Standards Writing Team, 2013, p. 4). They divided those progressions into grade level standards, and after revising, they published an updated series of documents to illustrate how the content standards follow a progression.
>
> Note that these documents are not themselves learning progressions, as we have defined them here. The authors point out that they "explain why standards are sequenced the way they are, point out cognitive difficulties and pedagogical solutions, and give more detail on particularly knotty areas of the mathematics" (Institute for Mathematics and Education, 2007). That is, they present related standards in sequences that correspond to the underlying progressions, rather than present the progressions themselves.

As she reads, she finds some compelling big ideas and starts making a list for herself. Some of the ideas she includes are ideas that, while not unfamiliar to her, are not really things she had thought about before:

- Measuring length can mean the length of an *object* or the length of a *space* (such as the distance between me and my friend or the difference between my height and my sister's height).
- Smaller units can be combined into larger units of units, such as a kilometer as 1,000 meters or a yard as 3 feet.

Other ideas she includes are ideas that she is familiar with, but she knows that they can be difficult for students to understand:

- The same object can have different measurements, depending on the measurement unit you use (for example, a line segment can be 3 inches, one fourth of a foot, or about 7.5 centimeters).
- Measurements are approximations. The level of precision you need for any measurement depends on the purpose for which you're measuring.
- Area is a measure of two-dimensional space. Although surface area is calculated on three-dimensional objects, it's still a measure of two-dimensional space.

She also includes ideas that seem basic, because she realizes they are fundamental to students' understanding. Without understanding these building block ideas about measurement, she knows that students could end up very lost:

- When you count units of measure, you're counting units of distance, space, or volume.
- When you combine measures with units, the units have to be identical to each other; they can't be different sizes (for example, 3 inches + 2 centimeters ≠ 5 units).

- Regions of area can be decomposed into smaller regions of area in different ways, without changing the total area of the figure.

She eventually ends up with a list of 20 big ideas. She decides to ask her colleague, Mr. Grafton, who also teaches seventh grade down the hallway, to take a look at her list and see what else occurs to him. He has several ideas to add, and they agree that between them, they might not have all the important ideas, but they do have a good place to start.

They decide to continue working together, to plan for their upcoming unit on surface area and volume. They meet the next day after school, and they try to organize the ideas into early understandings that students need to have about area, surface area and volume; intermediate understandings that would follow the early ones; and later understandings that follow the intermediate ones. As they sort, Ms. Lucenta realizes she finds it difficult to figure out how far back to go in listing important ideas. There are many ideas that seem so basic, but Mr. Grafton points out that if students never learned some of these basic ideas, that lack of understanding might show up in a variety of different kinds of mistakes, so he argues for including them.

They put each idea on a Post-It note and arrange them in different piles on the table in front of them. As they start arranging, Ms. Lucenta remarks that she sees some groupings that make sense to her. Some of the ideas seem to relate to a concept of *measure*—fundamental ideas about what it means to measure something. Mr. Grafton picks up on this, and he notes that other ideas seem to him to be about understanding *units* in measurement.

By the time they are done, they have arranged their ideas into early, intermediate, and later understandings (Figures 6.3–6.5), as well as

Figure 6.3   Early Understandings About Measures

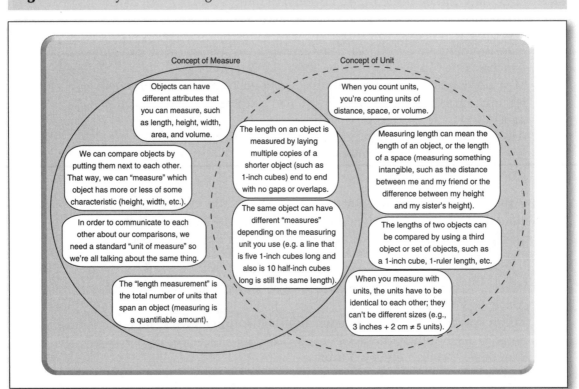

Figure 6.4   Intermediate Understandings About Measures

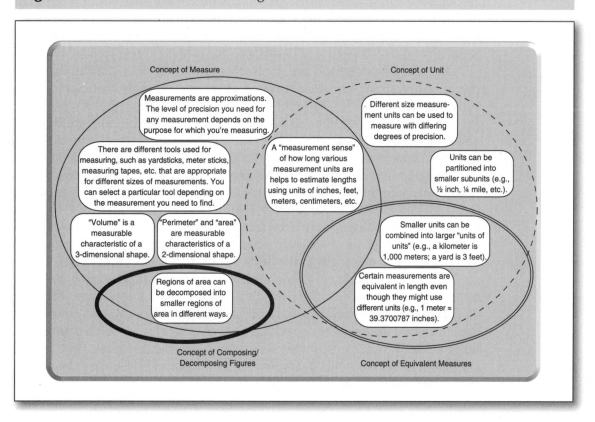

Figure 6.5   Later Understandings About Measures

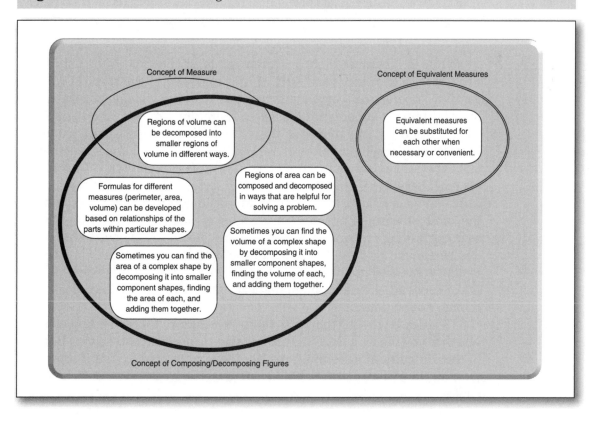

grouping ideas related to the concepts of *measure*, of *unit*, of *composing or decomposing figures*, and of *equivalent measures*. Some ideas sit in more than one category. No sequencing was done within the diagrams—Ms. Lucenta and Mr. Grafton will consider those kinds of connections among the ideas soon.

Once they have sorted the ideas, they put their Post-It notes on large pieces of newsprint and start drawing arrows between the various ideas to show which ideas lead to others. Mr. Grafton suggests that by the time they're done, he would like to be able to pick an idea in their progression and follow the arrows backward to see all the different prior ideas that feed into that chosen idea. They both agree that the early understandings are, in general, prerequisite ideas for understanding the intermediate ideas, which in turn are prerequisite ideas for the later understandings. However, when there are connections leading from one idea to another that they particularly want to highlight, they note that. By the time they have to wrap up, they have created the following road map. (See Figure 6.6.) Again, they agree it is not as complete as it could be, but it's a useful first draft.

Ms. Lucenta and Mr. Grafton's planning activity is providing them an opportunity to build on their existing mathematics knowledge, to think about certain ideas that they may not have thought about, and to connect certain mathematics ideas for which they may not have seen a connection before. This activity is one that we have done with teachers often in our professional development work, and from that experience, we provide our first recommendation.

> **Recommendation 1: Be strategic about whether you create a learning progression or use an existing one.**

While a fair amount of research exists in the field about students' mathematics learning, very little of it is currently in the form of existing learning progressions, and even fewer are readily available and usable by middle school mathematics teachers. Therefore, you may be tempted to create your own. We have done this with teachers who participated in our professional development, and we can say that it is a very worthwhile task. The conversations that are generated *in the process of creating the learning progression* have always, in our experience, been very rich mathematically and have felt worthwhile to all involved. For that reason, we feel that the value of learning progressions can lie as much in the creation of them as in the having of them once they're completed. However, it is difficult to describe many of these big ideas in the progression without having first read a great deal about what is known in the field about how students make sense of different mathematical ideas. Generating the ideas from scratch can be daunting.

This leads to our current recommendation to be strategic about whether you create a learning progression or use an existing one. Because of the difficulty in generating the big ideas from scratch, we recommend working from an existing learning progression when one exists, as a first choice.

Figure 6.6 A "Road Map" Learning Progression for Understandings About Measure

Early Understandings About Measures
Concept of Measure

Objects can have different attributes that you can measure, such as length, height, width, area, and volume.

When you count units, you're counting units of distance, space, or volume.

We can compare objects by putting them next to each other. That way, we can "measure" which object has more or less of some characteristic (height, width, etc.).

Measuring length can mean the length of an object, or the length of a space (measuring something intangible, such as the distance between me and my friend or the difference between my height and my sister's height).

In order to communicate to each other about our comparisons, we need a standard "unit of measure" so we're talking about the same thing.

The length on an object is measured by laying multiple copies of a shorter object (such as 1-inch cubes) end to end with no gaps or overlaps.

The lengths of two objects can be compared by using a third object or set of objects, such as a 1-inch cube, 1-ruler length, etc.

The same object can have different "measures" depending on the measuring unit you use (e.g., a line that is five 1-inch cubes long and also is 10 half-inch cubes long is still the same length).

When you measure with units, the units have to be identical to each other; they can't be different sizes (e.g., 3 inches + 2 cm ≠ 5 units).

The "length measurement" is the total number of units that span an object (measuring is a quantifiable amount).

Concept of Unit

Intermediate Understandings About Measures
Concept of Measure

Different size measurement units can be used to measure with differing degrees of precision.

Units can be partitioned into smaller subunits (e.g., ½ inch, ¼ mile, etc.).

Measurements are approximations. The level of precision you need for any measurement depends on the purpose for which you're measuring.

A "measurement sense" of how long various measurement units are helps to estimate lengths using units of inches, feet, meters, centimeters, etc.

There are different tools used for measuring, such as yardsticks, meter sticks, measuring tapes, etc. that are appropriate for different sizes of measurements. You can select a particular tool depending on the measurement you need to find.

"Perimeter" and "area" are measurable characteristics of a 2-dimensional shape.

"Volume" is a measurable characteristic of a 3-dimensional shape.

Regions of area can be decomposed into smaller regions of area in different ways.

Smaller units can be combined into larger "units of units" (e.g., a kilometer is 1,000 meters; a yard is 3 feet).

Certain measurements are equivalent in length even though they might use different units (e.g., 1 meter = 39.3700787 inches).

Concept of Unit

Concept of Equivalent Measures

Concept of Composing/Decomposing Figures

Concept of Equivalent Measures

Equivalent measures can be substituted for each other when necessary or convenient.

Later Understandings About Measures
Concept of Measure

Regions of volume can be decomposed into smaller regions of volume in different ways.

Regions of area can be composed and decomposed in ways that are helpful for solving a problem.

Formulas for different measures (perimeter, area, volume) can be developed based on relationships of the parts within particular shapes.

Sometimes you can find the volume of a complex shape by decomposing it into smaller component shapes, finding the volume of each, and adding them together.

Sometimes you can find the area of a complex shape by decomposing it into smaller component shapes, finding the area of each, and adding them together.

Concept of Composing/Decomposing Figures

203

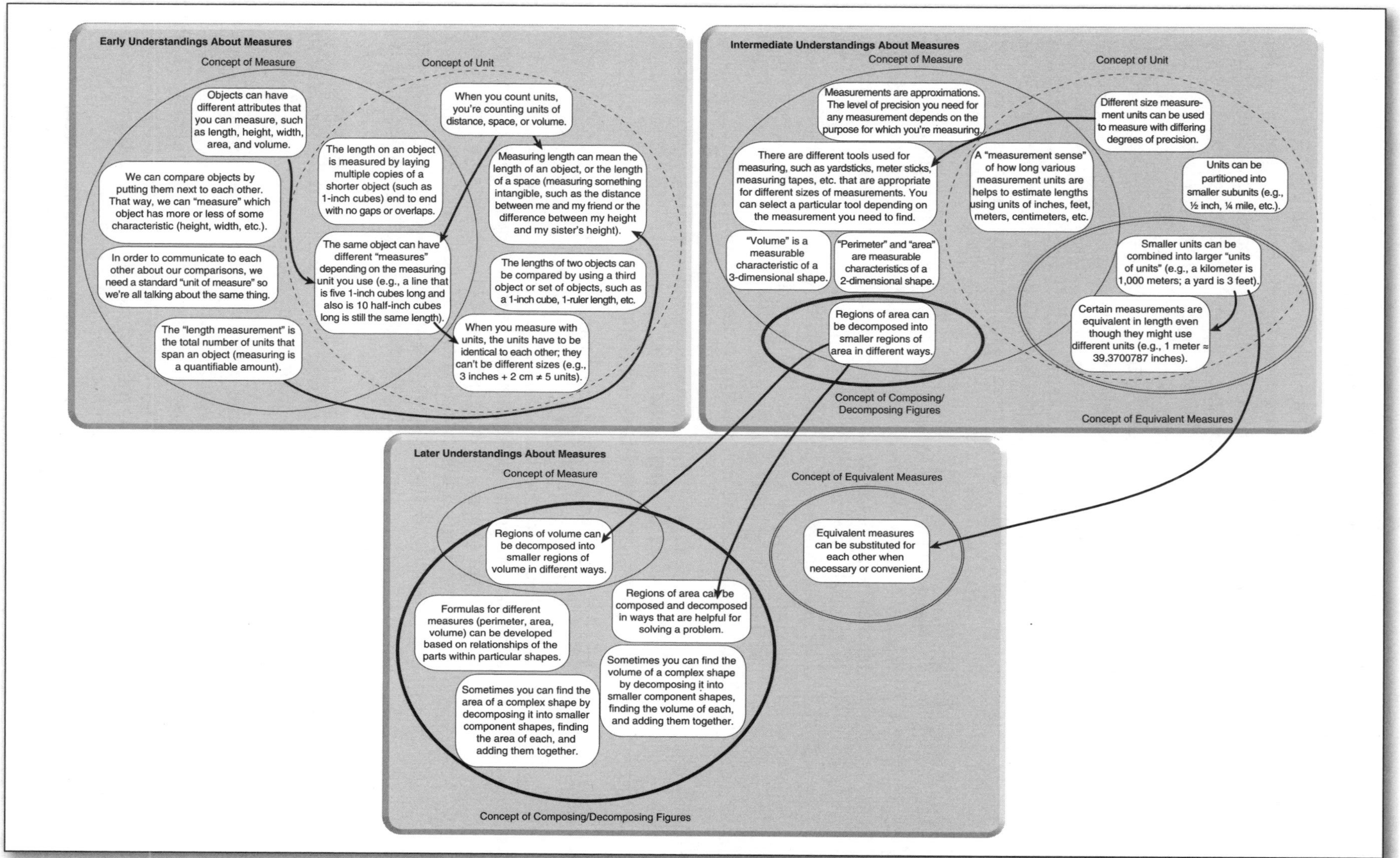

If and when you do decide to create your own, we have a few suggestions that we strongly encourage you to consider:

- Use research and other resources to help you. Read up on the ideas in the field before you come together to think about the details of the topics. A selection of articles and books that might be helpful is listed in the Resources section of this chapter. In particular, the book *Mathematics Curriculum Topic Study: Bridging the Gap Between Standards and Practice* (Keeley & Tobey, 2006), provides a process and set of tools that can help you better understand, access, and use mathematics education research.
- Pick a specific topic to work on. The more specific it is, the better you're able to focus your work. If you feel you're putting too much time in and the result is becoming unwieldy, try narrowing the topic.
- Start small—don't try to do all topics for the year. This can be difficult work, so take the time to work these in gradually, over multiple years.

Some teachers may wonder why they can't just use their teacher editions or teacher resources that come with their curriculum materials. While teacher materials with any set of curriculum materials do provide a wealth of valuable information, most tend to focus on *what* students will learn, not *how* the learning is expected to develop over time (whether over the course of a unit, a chapter, or several units or chapters). And as we said, if you can do the activity with at least one other person, the conversations you will have as you create this road map are invaluable.

Purpose 2: Using Learning Progressions to Plan Instruction

The following week, Ms. Lucenta and Mr. Grafton get together after school to start mapping out the unit they will teach on surface area and volume. They decide to use the learning progression they've created to identify some key concepts to address in the unit. They know they want to review the concept of area, as well as the use of area formulas for triangles, parallelograms, trapezoids, and regular polygons. In addition, they know they need to address the concept of surface area and how to find the surface area of various prisms and other, more complex polyhedra. However, they realize that this list of ideas focuses predominantly on skills, so they look at their progression to target some larger concepts as well.

Looking at their progression, they decide that the focus of their unit is most closely related to the following key concepts:

- Regions of area (and volume) can be decomposed into smaller regions of area (or volume) in different ways.
- Area is a measurable characteristic of a two-dimensional shape; volume is a measurable characteristic of a three-dimensional shape.
- Regions of area (or volume) can be composed and decomposed in ways that are helpful for solving a problem.
- Formulas for different measures (perimeter, area, volume) can be developed based on relationships of the parts within particular shapes.

As they identify this list of key concepts, several things jump out at Ms. Lucenta, and she comments to Mr. Grafton, "You know, for every thing we say here about area, we can say the same thing about volume. There's a real parallel between those two types of measures that seems apparent here. I bet we could build on that to help the students make connections between the two." She continues, saying, "And I'm noticing that the idea of composing and decomposing shapes and figures is coming up a lot. Makes me think we should really spend some time helping students understand the idea that you *can* compose and decompose figures, and helping them to build the visualization skills to do so." Mr. Grafton adds, "That's certainly going to help them as we work on surface area and when we try to find volume or surface area of complex figures."

They decide to break down the key ideas from the progression into a list of smaller, more bite-size ideas or subconcepts that could form the focus of a lesson or series of lessons in their unit. They also decide to note any necessary skills that go along with those ideas. We refer to the plan that Ms. Lucenta and Mr. Grafton created (Table 6.2) as a *unit progression*. The unit progression's ideas are listed in a rough order, with the understanding that the subconcepts will likely be addressed and revisited in different ways by multiple lessons over the course of the unit.

Table 6.2 Example Unit Progression

Smaller Subconcepts	Related Skills
You can decompose a shape into smaller component shapes to find the area of the original shape. You can do the same with three-dimensional figures.	• Subdivide a figure into smaller component shapes. • Use decomposition to find the area of two-dimensional shapes. • Use decomposition to find the volume of three-dimensional shapes.
Surface area is the sum of the areas of each of the two-dimensional faces of a three-dimensional figure. So to find surface area, you need to identify all the separate two-dimensional figures in the net of a three-dimensional figure.	• Subdivide a figure into smaller component shapes/create a net of a three-dimensional object. • Find the surface area of any prism. • Find the surface area of a complex three-dimensional polyhedron. • Distinguish between situations that call for area measure versus that call for surface area measure.
You can build formulas for surface area by combining area formulas you already know.	• Create a formula for surface area by putting together the relevant area formulas. • Find the surface area of any prism. • Find the surface area of a complex three-dimensional polyhedron.
You can develop volume formulas by decomposing the inside space of a figure into cubes or cross-sections/slices.	• Decompose figures into cubes, cross-sections, or other units of measure. • Find the volume of a figure. • Develop a volume formula for a figure.

Creating a Unit Progression to Clarify Concepts

Our next recommendation suggests using this idea of a unit progression as a way to clarify, for yourself, the concepts you want to emphasize and connect in the unit.

> **Recommendation 2: Create a unit progression to help define the portion of the learning progression you'll address in a unit of study and to clarify the important mathematics concepts that will be emphasized in your unit.**

Learning progressions span a period of learning that includes many grade levels and are written from an aerial view, as if you are viewing the development of your students' learning from far above; the overall path is clearer, but the details are hard to make out at that level. If you were to descend to a lower altitude, you could get a clearer picture of the details within a certain portion of that path. Creating a unit progression does precisely that, and it can be a valuable first step when planning a unit. Note that a unit progression is *not a pacing guide* for a unit. Listing four subconcepts does not imply that there are four lessons that take place over four days. Instead, a subconcept may be addressed over several days and several separate lessons.

The following steps summarize what Ms. Lucenta and Mr. Grafton did, and following them can help you create such a unit progression:

1. **Define the portion of the learning progression, if you have one, that will be addressed in your unit.**

   Because learning progressions do not specify particular grade levels or units of study, you will need to decide what portion of the learning progression you'll try to address in a particular unit of study. Depending on whether your unit is introductory to a topic or reviewing and pulling together many ideas, you may choose to use only a few or several of the bigger concepts from a learning progression. You may revise the start and end points of your unit as you work through these steps, so don't feel tied to your initial decision. When it isn't possible to start with a learning progression, many of the teachers we worked with for this first step considered the mathematical big ideas of their unit and how these ideas connected to big ideas in related previous units and subsequent units.

2. **Identify important smaller concepts that are part of understanding the larger concepts stated in the learning progression.**

   As you think about each of the larger key ideas, you can break each one down into a short list of smaller-grain-sized concepts that are part of understanding the larger key idea. For example, a larger idea from Figure 6.4, "'Perimeter' and 'area' are measurable characteristics of a 2-dimensional shape," could be broken down into smaller concepts including, but not limited to, the following:

- Perimeter is a measure of the distance traced around the outside of a two-dimensional shape.
- Area is a measure of space covered within the interior of a shape.
- Because perimeter measures length (one dimension) and area measures space covered (two dimensions), they each use different units of measure.

3. **Put those smaller concepts in a mathematically logical order that defines how your unit will build across different lessons.**

   The process of identifying the smaller concepts and putting them in an order that makes sense results in a set of mathematical headlines for your unit. This process should allow you in the end to be able to describe how the unit starts and ends, how it builds, and why it builds in that particular manner. (This is a good step for firming up the beginning and end of your unit.)

4. **Establish connections between and among concepts in the unit that may not have been apparent initially.**

   Often, the process of creating this unit progression can illuminate connections between and among concepts that you may not have noticed or that you now view in a new way; this happened for Ms. Lucenta when she noticed how two-dimensional area and three-dimensional volume had many parallel concepts. Establishing these connections can help you foreshadow them in your lessons and encourage students to make those connections as well.

Most teachers work from curriculum materials that already have chapters or units created for them, so it may be tempting to just skip this process of generating a unit progression. However, if you have some freedom to determine the order and timing of your lessons, you might find that in creating the unit progression, you will want to try a different sequence of lessons than the one that exists in your materials. One caution in doing so, though, is to be sure to work through the activities and problems yourself, in the order you want to do them. If your materials were developed to build concepts in a specific way, you want to be aware of how that development takes place and not inadvertently overlook something important as you reorder the lessons. Even if you find something that doesn't fit, you can still reorder them, but you may want to skip particular problems or find a replacement task to better meet your needs.

If, on the other hand, you don't have this freedom—for example, if you work in a district that uses a pacing guide—creating a unit progression using the existing order can still provide useful insights into the connections among the topics. You might discover a way to emphasize early concepts that will provide stronger gains later in the unit, for example; and having an idea of how the concepts build might allow you to take advantage of insightful remarks—or budding misconceptions—that your students make. In addition, many materials do not include concepts and subconcepts, so thinking about these across the unit will help you write your learning intentions (see the next section).

> ## TEACHERS' VOICES
>
> I was beginning a lesson on seeing fractions as division problems. In my warm-up problem, I discovered that many of the students, who were still new to me, were having difficulty understanding improper fractions. I hadn't considered that their understanding of improper fractions might be a barrier!
>
> Next time, I will do a unit progression first, to help me think about these prerequisites and make these connections. I've found unit maps helpful in other units, and I know it would have been helpful to me here, too.

 Planning Resources: Go to Resources.Corwin.com/ CreightonMathFormativeAssessment to access these resources.

 Building a Unit Progression. This printable file provides guidelines for developing unit progressions.

 Unit Progression Builder. This printable file uses the process from "Building a Unit Progression" above to help you build a progression for any unit you are planning.

Using a Unit Progression to Write Learning Intentions

After completing the unit progression with Mr. Grafton, Ms. Lucenta began planning her lessons for the next week. For each of the subconcepts in the unit progression, she thought about how many lessons she would need to teach that idea and what would be the primary goal of each lesson. From that, she generated a set of learning intentions to match the lessons she imagined. In some cases, she felt that the subconcept was already stated in a way that seemed like a good learning intention, though she knew she would need to clarify the word *decompose*.

Subconcept From Unit Progression	Learning Intention/Success Criteria
You can decompose a shape into smaller component shapes to find the area of the original shape.	*You can decompose a shape into smaller shapes to find the area of the original shape.*

Subconcept From Unit Progression	Learning Intention/Success Criteria
	• I can decompose a figure into smaller shapes that form it. • I can use the areas of those shapes to find the area of the complete shape. • I can explain how I decide to decompose the shape into smaller pieces.

In other cases, she felt the subconcept needed to be broken down into several different lessons, each with a slightly different focus:

Subconcept From Unit Progression	Learning Intention/Success Criteria
Surface area is the sum of the areas of each of the two-dimensional faces of a three-dimensional figure. So to find surface area, you need to identify all the separate two-dimensional figures in the net of a three-dimensional figure.	*(Lesson 1) Any polyhedron can be represented as a flat net.* • I can create a net of a polyhedron. • I can sketch or build a polyhedron from a net. • I can explain how I know whether a net is a correct two-dimensional representation of a given polyhedron. *(Lesson 2) The net of a three-dimensional object can help us find the object's surface area.* • Using a three-dimensional object, I can point out what surface area is. • I can explain how I use what I know about the area of a two-dimensional object to find surface area of a three-dimensional object.

Ms. Lucenta's process embodies our third recommendation.

> **Recommendation 3: Use a unit progression to support your writing of your learning intentions.**

A unit progression serves as a magnification of one part of your learning progression. It makes clearer the relationship between the learning that has to take place in your classroom and the larger ideas articulated in the learning progression. The subconcepts can form the basis of individual learning intentions that you develop for a lesson, and they can therefore significantly ease the process of writing learning intentions. Sometimes the subconcepts themselves might be just the right learning intention you need for a lesson; other times, your learning intention might grow out of one of the subconcepts.

Planning Resource: Go to Resources.Corwin.com/ CreightonMathFormativeAssessment to access this resource.

Guidelines for Writing LI and SC. This printable file provides guidelines for using unit progressions to write effective learning intentions (LI) and success criteria (SC).

Reference Resource: Go to Resources.Corwin.com/ CreightonMathFormativeAssessment to access this resource.

Chapter 6 Summary Cards. This printable file provides Summary Cards that include the steps described for building a unit progression and for using a unit progression to develop LI and SC planning guidelines.

Learning Resource: Go to Resources.Corwin.com/ CreightonMathFormativeAssessment to access this resource.

Sample Unit Progressions. These example unit progressions were created collaboratively with teachers from our professional development work.

Purpose 3: Using Learning Progressions to Help You Diagnose Student Difficulties

The learning progression can also help you diagnose gaps in students' understanding that may be hindering them from reaching the learning intention. Using a learning progression in this way can be compared to analyzing and determining the quality of a building structure. When you are buying a house, the house you love with a beautiful exterior may turn out to have cracks in the walls on the second floor. These cracks can be due to floor supports that are insufficient as the house settles over time, and further study can reveal that the floor supports are insufficient because the foundation is crumbling and settling unevenly. Knowing to look for a solid foundation as a key basis for supporting the rest of the house is important when judging the structural strength of a house. In the same way, analyzing and determining the nature of a student's difficulty may require peeling back some layers of understanding to determine what the possible root cause is of a particular difficulty. Trying to "follow the arrows backward," as Mr. Grafton wanted to do—whether with a learning progression or by consulting articles and books about how students learn about a particular concept—can point to other fundamental prerequisite concepts whose lack may prove to be the cause of a student's difficulty.

Let's return to Ms. Lucenta and Mr. Grafton to see an illustration of this.

Mr. Grafton gave his students broken rulers (Figure 6.7), portions of a ruler cut out of paper strips that they could use to measure the dimensions of a photograph. The task is designed to elicit two common misconceptions that students have around linear measurement: They count measurement line markings rather than the spaces between the markings (a concept of measurement of length as a measure of the distance *between* two points), and they do not keep the unit of measure consistent. Students who have these misconceptions will make one or more of the following mistakes:

- They read off the number on the ruler that appears at the end of the measurement, disregarding what number on the ruler they have started to measure from.
- They sometimes count the number of lines marked with whole numbers on the ruler, rather than count the spaces between the markings.
- When more than one broken ruler is needed, they don't overlap them to line up the ruler markings properly and maintain a constant unit of measure; instead they place the rulers end to end, which introduces partial units that they might count as one whole unit or not at all.

Figure 6.7 Broken Ruler Segments

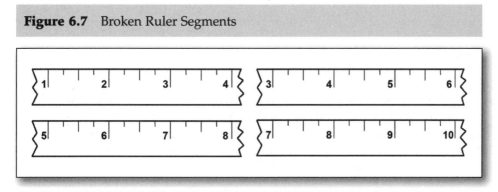

Source: © EDC.

Mr. Grafton walked into Ms. Lucenta's room during their prep period to show her a piece of student work that he had collected. "This is so interesting!" he said. "Or maybe depressing. Or maybe both. I gave my third-period class the task from the article we read the other day that targets student misconceptions around measurement. And look what I got back." (See Figure 6.8.)

Ms. Lucenta studied the work for a moment, then remarked, "Wow. There's certainly more going on here than not being able to read a ruler!" She pulled out the piece of paper on which they had sketched out their learning progression. "I remember a couple ideas that came up when we were creating this, which I hadn't thought about before, really, because I just considered them elementary school topics." She pointed to two of the key ideas in the "early" part of the progression:

- When you count units, you're counting units of distance, space, or volume.
- The length of an object is measured by laying multiple copies of a shorter object (such as 1-inch cubes) end to end with no gaps or overlaps.

Figure 6.8   Broken Ruler Student Work

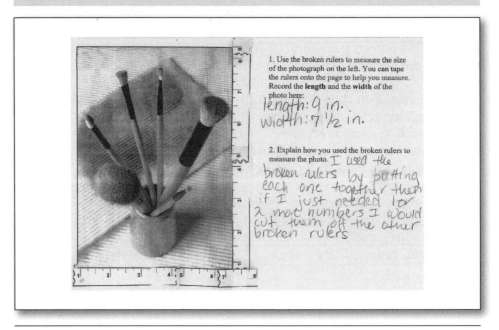

Source: © EDC.

"I wonder if your student never really solidified these ideas way back whenever they were first introduced to her," she added. Mr. Grafton agreed and headed back to his room, thinking about how to find out more from his student about what she understood about measurement.

Mr. Grafton's experience illustrates our fourth recommendation.

> **Recommendation 4:** When deciding on appropriate responsive actions, consider the learning progression (or at least consider information about how the relevant concepts build) to help you determine the nature of any gaps in students' understanding.

Having a solid knowledge of a learning progression in the back of your mind can be invaluable in helping you interpret the various evidence you collect about your students' mathematical reasoning. A medical doctor uses background knowledge about how the body works to identify different possible causes of your symptoms and (in most cases) has enough resources and tools to narrow down what medical condition is the root cause of your symptoms; similarly, learning progressions can help you to diagnose whether key understandings are missing and then to treat that gap in understanding with targeted instruction. We certainly want our doctors to be very well informed so they do not misdiagnose an ailment; otherwise we spend time and aggravation trying to solve the wrong problem. The evidence that suggests a student needs assistance may be masking the real source of the difficulty, but progressions give you information to more accurately pinpoint the nature of the gap in understanding, and they help to focus your instructional efforts in the right place. For example, armed with the key ideas Ms. Lucenta pointed out, Mr. Grafton could follow up with his student to gather more evidence to determine if one of those ideas is causing the student's difficulty.

Learning progressions can also support your instructional decisions about next steps. They allow you to see how your current content fits within a larger context of learning—both what came before and what will follow from the content you're teaching now. This knowledge of how the current content is laying the foundation for students to develop more sophisticated understanding later can help you maximize any instructional opportunities to foreshadow those future concepts. Returning to the house analogy, imagine a long-term renovation project, with several stages. While a certain amount of unfinished clutter is acceptable, you want each stage to look complete—but it must be based on the existing foundation *and* be constructed so that the next stage can be constructed properly as well. In the same way, knowledge of a learning progression allows you to both attend to the current learning as it is evolving and also be simultaneously aware of how the content you're teaching develops from what students have learned previously and how the current content is related to subsequent learning. For example, teaching whole-number division might build on the area models that are used for teaching multiplication; this model could then be extended to division with fractions. Or, for a tighter time frame, knowing that a topic will be brought up in the next couple of lessons might suggest that a comment or question made by a student could be highlighted (as foreshadowing) but not delved into yet because the questions will be answered soon.

USING LEARNING PROGRESSIONS TO HELP DEVELOP STUDENT OWNERSHIP AND INVOLVEMENT

Learning progressions themselves are largely for the use of the teacher in planning and in interpreting evidence, and they are often not appropriate as tools to share with students. However, some teachers have found benefits in sharing information gleaned from a learning progression with their students.

Sometimes this sharing can involve making public the development of skills or understandings that is intended to occur over the course of a unit. As an example, consider a seventh-grade classroom in which a regular-education teacher was team-teaching with a special education teacher, and a large portion of the students had Individualized Education Programs (IEPs). The class was learning about ratios, and they had discussed different ways in which you could compare ratios. These strategies included making tables, using a scale factor, finding a unit rate, and solving a proportion. The teachers arranged the list of strategies from most accessible to most abstract, and they had posted the list publicly for students to reference, building on the list as different strategies were discussed. As they introduced the lesson for the day, they asked students to think about which strategy they were currently most comfortable with, and they explained that today the students would focus on getting more proficient with the next more-challenging strategy. They divided the students into small groups according to the strategy they were working on (some were allowed to self-identify, others were assigned by the teachers), and the teachers worked with different groups during the lesson. In this case, sharing and discussing the list of strategies with students allowed some students to do some self-assessment of their own placement within the progression of skills and to make individual decisions about where to focus their efforts. Making public the list of different strategies gave all students a clear picture of how they needed to progress.

Sometimes this sharing can focus specifically on sharing an overview of the unit progression with students, to use the unit progression as a road map through the unit. At various points throughout the unit, it can be useful to point out to students where the current lesson is focused—an opportunity to pause and note that "You are here." Consider an example of a sixth-grade teacher who posted the subconcepts from her unit progression for her class. She used them at the start of the unit to talk to the students about the focus and direction of the unit, and she asked students to self-assess which of the concepts they felt they had some familiarity with and which were brand new to them. She revisited the list of subconcepts at various points during the unit to pause and reflect on how the current lesson related to their overall path through the unit and to discuss how the lesson connected to what came before and to what was coming next. She also used the list to reflect and summarize with students at the end of the unit. The use of a unit progression as a tool at before, during, and after points of teaching the unit parallels the same kind of use of learning intentions and success criteria throughout a lesson. This parallel is not accidental; unit progressions serve to identify and articulate the important mathematics learning at different grain sizes, just as learning intentions and success criteria do.

■ CONCLUSION

Although there are numerous tools available to teachers to identify, map, and organize mathematics learning goals and standards, learning progressions stand out as a tool with a different purpose. A well-constructed mathematics learning progression is a vital resource for teachers to deepen their mathematics knowledge, to better understand the nature of students' mathematics learning, to more accurately diagnose the cause of students' mathematics difficulties, and to create appropriate learning goals. Learning progressions take many forms. While there seems to be no consensus in the literature about what they look like or what breadth or depth of content they cover, one thing remains true of all learning progressions: They lay out, in some way, a clear path of how learning develops from more rudimentary to more sophisticated understanding for a particular topic over some time.

Knowledge of a mathematics learning progression for any topic influences the creation of learning intentions, the interpretation of evidence in identifying gaps of understanding, and the nature of different responsive actions. Because of its pervasive influence on all the different critical aspects of formative assessment, it is an important supporting aspect. In the next chapter, we consider the other supporting aspect: the classroom environment.

Recommendations in This Chapter

1. Be strategic about whether you create a learning progression or use an existing one.

2. Create a unit progression to help define the portion of the learning progression you'll address in a unit of study and to clarify the important mathematics concepts that will be emphasized in your unit.

3. Use a unit progression to support your writing of your learning intentions.

4. When deciding on appropriate responsive actions, consider the learning progression (or at least consider information about how the relevant concepts build) to help you determine the nature of any gaps in students' understanding.

Reference Resources: Go to Resources.Corwin.com/ CreightonMathFormativeAssessment to access these resources.

Chapter 6 Summary Cards. This printable file provides Summary Cards that include recommendations in this chapter.

Formative Assessment Recommendations. This printable file provides a summary of all the recommendations in this book.

RESOURCES ■

The following sections present some resources that can help you implement the recommendations. Each of these resources is referenced in earlier sections of the chapter, but here we provide a consolidated list. All resources can be found at Resources.Corwin.com/CreightonMath FormativeAssessment.

Learning Resources

These resources support your learning about formative assessment practices related to using mathematics learning progressions.

- **Sample Unit Progressions.** We created the progressions collaboratively with teachers in our professional development program.

- **Books and Articles List** includes a list of potential resources for learning progressions and for creating learning progressions and unit progressions.

Reference Resources

These resources summarize key ideas about formative assessment practices related to using learning progressions:

- **Chapter 6 Summary Cards** provides index-size Summary Cards for the following:
  - **Building a Unit Progression** summarizes the steps involved in building a unit progression.
  - **Using a Unit Progression to Write LI and SC** summarizes the steps described in the guideline document in Planning Resources.
  - **Recommendations for Learning Progressions** includes all the recommendations from this chapter.
- **Formative Assessment Recommendations** is a printable PDF that includes all the recommendations from this book.

Planning Resources

These resources support your planning process:

- **Building a Unit Progression** includes a step-by-step process for building a unit progression.
- **Unit Progression Builder** is an interactive web page that uses the process in the "Building a Unit Progression" resource to help you build a progression for any unit you are planning.
- **Guidelines for Writing LI and SC** includes a step-by-step process for writing learning intentions and success criteria; a unit-level progression can help inform the process.

7

Establishing a Classroom Environment

We've now looked at all of the various aspects of formative assessment but one: the classroom environment. With what we hope is a clear and complete vision of effective formative assessment practices, you can turn your attention to a classroom environment that supports it. Any teacher understands how critical his or her classroom environment is to how learning takes place. Charlotte Danielson (2007) wrote,

> *The classroom environment is a critical aspect of a teacher's skill in promoting learning. Students can't concentrate on the academic content if they don't feel comfortable in the classroom. If the atmosphere is negative, if students fear ridicule, if the environment is chaotic, no one—neither students nor teacher—can focus on learning. (p. 64)*

More specific to formative assessment, Margaret Heritage (2010) wrote,

> *The whole process of formative assessment depends on a classroom culture where students feel safe to say they do not understand something and give and receive constructive feedback from peers. Teachers must establish a classroom culture characterized by a recognition and appreciation of individual differences. (p. 15)*

These quotes both speak to the need for a classroom environment that incorporates a social culture that recognizes and supports the risks needed to move learning forward—that is, a classroom environment that allows you to embed formative assessment practices as a daily process.

While a lot can be said about classroom environment, we focus in this chapter on what it is about the classroom environment that serves formative assessment practices. First, though, we need to establish more clearly what we mean by classroom environment. Then we look at the elements of the classroom environment and make some recommendations to help you establish an environment conducive to formative assessment practices.

■ ELEMENTS OF THE CLASSROOM ENVIRONMENT

On dictionary.com, the first definition of the word *environment* is "the aggregate of surrounding things, conditions, or influences." Determining what is the classroom environment for a particular classroom can be particularly elusive to pin down and complicated to discuss, as it comprises such intangible and interrelated pieces as conditions and influences. However, since the environment sets a tone and a context for all learning that takes place, examining it is vital. It defines the relationship between the teacher and students, and it's a direct reflection of any teacher's personality, teaching style, and beliefs and assumptions about teaching and learning. For our purposes, we consider the classroom "aggregate of surrounding things, conditions, or influences" using three viewpoints: the social cultural environment, the instructional environment, and the physical environment. Let's look at how all three of these have an influence on your implementation of formative assessment practices.

- **The social/cultural environment is about how you and your students interact with each other.**

  What are the guidelines for how students treat each other in class? What can you and your students collectively do so everyone feels safe in class to share their ideas? In what ways can you encourage your students to stretch their thinking and take intellectual risks? Answers to questions like these that spell out the norms and guidelines for behavior and interpersonal interactions are what define the social and cultural environment in a classroom. Although these norms and guidelines often include necessary rules about physical safety, in this chapter, we look at the social and cultural environment primarily in terms of establishing social norms that promote intellectual safety and curiosity, so that students feel comfortable participating in formative assessment practices.

- **The instructional environment is about how you frame your instruction.**

  What kinds of math tasks are a good fit for providing opportunities for your students to meet the success criteria? To what extent and in what ways are students encouraged to share their thinking publicly? What's the right amount of scaffolding to provide students during a task? How much of the intellectual work in class are your students doing? All these questions are part of considering the instructional environment in your classroom. The kinds of math tasks students are given can either promote or hinder opportunities for students to make their thinking visible. A teacher can also optimize the learning for students by giving students every viable opportunity to do their own thinking—not just when solving problems but also in explaining concepts underlying a procedure, in making connections between two related ideas during a class discussion, and in pausing to summarize the important ideas in a lesson so far. In this chapter, we discuss the instructional environment in terms of framing instruction to encourage and make visible students' thinking around learning intentions and success criteria and to optimize learning to reach the learning intention.

- **The physical environment is about the classroom space and the materials available to your students.**

  How can you arrange the setup of your classroom to support using formative assessment practices? How can you best make available the various tools and resources your students need to become more proactive participants in the learning process? Answers to these and other questions about the physical space of your classroom define the physical environment. One kind of resource to consider is the physical materials—paper, calculators, manipulatives, and any other materials students might need, in order to make them easily accessible to students when needed. A second kind of resource is information, such as the learning intention and success criteria, or examples of work that meets the success criteria. A third kind of resource is peers. All teachers have deliberate arrangements of seating in their classroom, and we'll discuss how these choices of physical desk or table arrangements make peers more or less accessible to each other as a resource during learning. In this chapter, we look at the physical environment in terms of keeping accessible these resources, all of which support formative assessment practices.

All three of these elements of a classroom environment—social/cultural, instructional, and physical—have an impact on students' understanding and use of the learning intention and success criteria, on their ability to generate evidence of their learning and to evaluate their learning against success criteria, and on their ability to provide and respond to formative feedback. This affects both student self-regulation and the formative assessment process, as a whole.

Setting up a classroom environment conducive to formative assessment practices is fundamental to students being able to complete the steps in Thinking Like a Self-Regulating Learner (Figure 7.1).

Figure 7.1   Thinking Like a Self-Regulating Learner

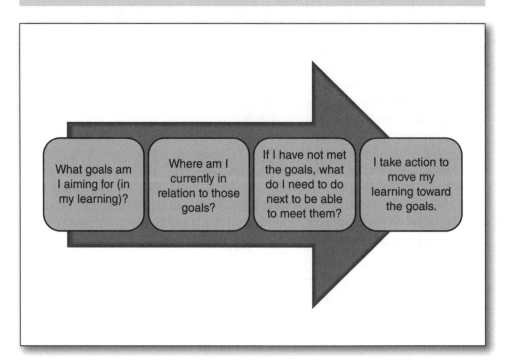

Because the environment sets a context in which all formative assessment practices are used, we include (at last!) this final supporting aspect in the Formative Assessment Cycle as an umbrella over the entire cycle. See Figure 7.2. As you'll see, teachers and students must work together to establish this classroom environment.

Example: For her lesson on relating number and visual patterns, Ms. Jenkins has chosen rich instructional tasks that require her students to stretch their thinking and make connections on their own. She uses the Exploratory Tasks to provide opportunities for students to discuss their thinking with each other, and the discussion between tasks allows different approaches to be heard.

Her students have learned that when they write something on their whiteboard, as they do in the warm-up activity, they should not change their answers when they see others with different answers. The classroom culture supports open sharing with respect for taking risks, so students know they can participate without fear of feeling ridiculed. This allows Ms. Jenkins to get a better sense of how students are thinking about the connection between the visual and numeric patterns (the learning intention) than she might if they were reluctant to share ideas that might be wrong.

As they work in pairs on the Exploratory Tasks, they have access to many resources—not only graph paper and plastic tiles that they can use to reconstruct the visual patterns if they like but also each other.

Figure 7.2 The Formative Assessment Cycle

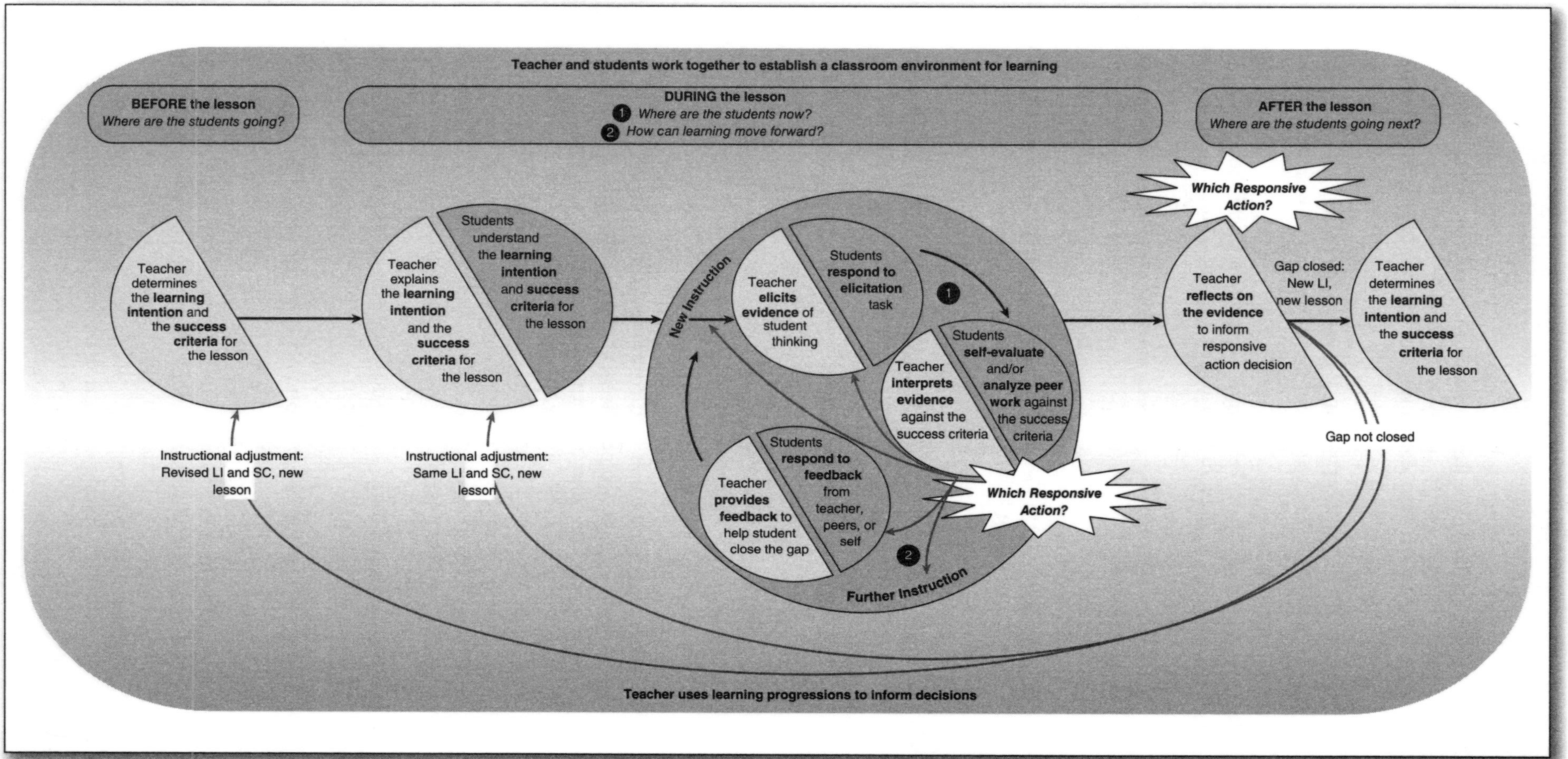

Teacher and students work together to establish a classroom environment for learning

BEFORE the lesson
Where are the students going?

DURING the lesson
1 Where are the students now?
2 How can learning move forward?

AFTER the lesson
Where are the students going next?

Teacher determines the learning intention and the success criteria for the lesson

Teacher explains the learning intention and the success criteria for the lesson

Students understand the learning intention and success criteria for the lesson

New Instruction

Teacher elicits evidence of student thinking

Students respond to elicitation task

Students self-evaluate and/or analyze peer work against the success criteria

Teacher interprets evidence against the success criteria

Which Responsive Action?

Further Instruction

Teacher provides feedback to help student close the gap

Students respond to feedback from teacher, peers, or self

Which Responsive Action?

Teacher reflects on the evidence to inform responsive action decision

Gap closed: New LI, new lesson

Teacher determines the learning intention and the success criteria for the lesson

Gap not closed

Instructional adjustment: Revised LI and SC, new lesson

Instructional adjustment: Same LI and SC, new lesson

Teacher uses learning progressions to inform decisions

221

Reference Resources: These resources are included in Chapter 8; however, you may want to go to Resources.Corwin.com/ CreightonMathFormativeAssessment to take a look now. Look for Chapter 8 Resources.

 Formative Assessment Cycle. This is a color version of the completed Formative Assessment Cycle.

 Formative Assessment Cycle Video. This is an audiovisual tour of the Formative Assessment Cycle that we are building throughout this book.

In the following sections, we discuss specific ways in which each of these elements of the classroom environment can support formative assessment, including both what the teacher can do and what students can do.

■ THE SOCIAL AND CULTURAL ENVIRONMENT: PROMOTING INTELLECTUAL SAFETY AND CURIOSITY

As you probably saw in Chapter 5, "Developing Student Ownership and Involvement in Your Students," one way to help students begin to learn the complex skills required for self-regulation is by practicing them with peers' thinking and work. Evaluating where a peer's work is in relation to the learning intention, assessing what the peer needs to do to be able to meet the learning intention, and communicating that information to a peer gives students a chance to practice this train of thought before applying it to their own work. The need to communicate this kind of information to a peer in a supportive and effective way points to the importance of establishing a particular social and cultural environment conducive to this kind of exchange.

Many of the teachers in our professional development work have talked about how they had already helped their students establish rules about responding to each other in class. They talk with their students about respect for each other's ideas, about promoting risk taking to put new ideas out there for discussion, and about how to be considerate when responding to other people's ideas. Closely linked to these are a classroom's approach to appropriate communication, such as focusing on an idea rather than the person and considering tone of voice. In some cases, classroom norms may even be focused on helping students understand how to interact in a conversation.

After focusing on how their particular norms connect to the formative assessment practices described in previous chapters, the teachers tweaked them slightly to better support their students. As with other formative assessment practices, they found they needed to help students learn to use the norms, by being explicit about the expectations for attending to these norms and then modeling what the norms look and sound like.

TEACHERS' VOICES

The staff at my school read the book *Knee to Knee, Eye to Eye* and incorporated the strategies into the classroom. The premise of this book is that students do not know the fundamentals of holding a conversation, and they need to be taught how to have a rich conversation through modeling and explicit steps. First, I showed a video about students having a book conversation, and half my students focused in on body language, and half my students focused in on the actual conversation. After, we brainstormed as a class how to initiate and maintain a conversation. This included focusing on one partner being the active listener and using body language to signify that he or she was listening. We also talked about the tone of voice and showing a respect for another's opinion. Students generated multiple ideas, and they practiced implementing the tips. Fostering more peer-to-peer interactions and discussing what a good conversation looks like sets the students up for success. It opens the door to take risks.

For the rest of this section, we present some examples of norms that teachers developed for use with students to support formative assessment and some recommendations for practices to support these norms. A formative assessment classroom is characterized by both opportunity and responsibility to think and act, so these sample norms focus both on providing such opportunities to students and on students taking responsibility to take advantage of those opportunities. Most of these norms are related to something we call intellectual safety and risk-taking, and there is one related to intellectual curiosity. For each sample norm, we recommend specific things related to the norm that you can do to develop your classroom environment to support formative assessment practices.

Intellectual Safety and Risk Taking

Students need to feel safe as they engage in classroom discussion and solve problems together, so you can elicit evidence of their thinking. This sense of safety for students includes a surety that they will not be ridiculed, that their thinking—whether correct or incorrect—will be considered, and that discussions about their own or others' thinking will be conducted with respect for the ideas and for the people offering them. This intellectual safety gives students permission to take risks with their thinking, so that they—and their peers—may learn from any mistakes they make. Let's look first at some sample norms that can support formative assessment practices, and then we can present some recommendations to help you establish them.

> **Sample Norm 1:** The classroom is a safe place for thinking and learning. Everyone has the opportunity to discuss their own and others' thinking with respect for the ideas and for the people offering them. All students also have the responsibility to be respectful of others and their ideas.

This norm establishes that intellectual safety is important. For students to learn, they all need to take risks in sharing their thinking, even when that thinking may contain some mathematical flaws or may not yet be fully developed. Making public their mistakes can feel very threatening, particularly at the middle school level, and many tasks to elicit evidence and provide feedback require them to reveal their thinking to their peers. The way in which this kind of flawed or undeveloped thinking is addressed and made part of the public conversation in class can make or break a student's willingness to engage in the conversation. While students are still learning a concept, the presumption should be that their thinking and understanding is still developing and therefore will have some flaws or areas that can be improved; unflawed thinking and perfect expression of ideas are never the goals until late stages of learning. When *all* thinking is treated as contributions to the collective understanding that the class is developing during a lesson, a student's sense of safety is reinforced, making it more acceptable to offer other ideas in the future.

> **Sample Norm 2:** Everyone in the class needs to participate. Everyone needs the opportunity to add something to whole-class work to contribute to the learning of others, and everyone has the responsibility to look for times in which they can contribute productively.

A large part of this norm rests on the previous norm, establishing a sense of intellectual safety in class. As students learn to trust this safety, the expectation that they will, in fact, take the leap of putting their thinking out for public discussion helps them not only to take ownership of their own learning but also to contribute to the learning of the class so that all may reach the learning intention.

> **Sample Norm 3:** Everyone in class should have the opportunity to think and process information. Everyone also has the responsibility to allow more time to those who need it.

Even when a teacher asks great questions, opportunities are lost when students do not have time to think about and respond to them. For example, Mr. Hughes poses many questions to his students during the course of a lesson, and he is particularly adept at asking questions that prompt some higher-order thinking. When his class was working on understanding the meaning of the slope-intercept form of a linear equation, $y = mx + b$, he asked questions to prompt students to describe the relationship between the mathematics symbols and the context of the problem. His questions included "You're right: b is the y-intercept, but what does that mean in terms of this problem?" and "So if I know m is 3, what does that tell me about the two things we're comparing?" However, in spite of asking thought-provoking questions whose answers could have both prompted good classroom discussion and given him good information about what students understood, he frequently went on immediately to answer his own questions. After a number of lessons in which this happened, the only students who spoke up to try to answer his questions were his strongest students, who could provide an immediate answer. Many of the students were unable to keep up during the class, and they had to catch up when they were working individually or in small groups.

Asking good questions but not allowing adequate time for students to think about answers defeats the effect of those good questions. It takes time for students to understand the question—especially if English is not their first language—and then it takes more time for students to formulate a response. Slowing down and providing this time not only allows more students to keep up with the instruction but also—and just as importantly—communicates to students that they really are expected to formulate an answer to the question.

For some students, the difficulty here is not in trying to keep up but rather in getting to a response quickly and wanting to respond as soon as possible. These students may see providing a quick answer as a helpful contribution to the class, without realizing that it's more helpful to the other students if they can reach the conclusion on their own. The second part of this norm, the responsibility to allow others time, may surprise these students.

With these and possibly other norms in place and enforced, your classroom environment can encourage the intellectual safety and risk taking that is needed for formative assessment practices to have their full impact. The following recommendations will help you establish these norms.

> **Recommendation 1: Be explicit about what it means to share your thinking and about why it's important.**

This recommendation helps establish Norm 1 (making the classroom a safe place) and Norm 2 (everyone needs to contribute). An important first step for students to take part in formative assessment is for them to provide their thinking, but many students don't know why they should or even what it means to share their thinking.

You'll find it helpful to talk explicitly with your students about the importance of knowing how a student is thinking about the mathematics—both so you (and they) can compare that thinking to the success criteria, and so they can learn from each other. If they are unaccustomed to being asked to explain their thinking, they will often assume that you are asking because their answer is incorrect, leading some students to just shut down and others to change their answer, rather than simply explaining their line of reasoning. It will take time, but their responses will improve as you reassure them that you're interested in their thinking, whether or not their answers are correct.

Providing them with structures and modeling such sharing will also help them to improve their responses. See Chapter 3, "Gathering, Interpreting, and Acting on Evidence," and Chapter 5, "Developing Student Ownership and Involvement in Your Students," for more about ways to help students share their thinking.

When students are engaging in any kind of cooperative learning activity, I've found that giving them explicit language directions and examples is incredibly helpful for them to be able to communicate with each other and to relay academic/content information. For example, when asking students to solve a division of fractions problem together, I model and post sentence starters, such as "I set up my equation like this, _____ . . ." or "First, I wrote _____ because _____. Then, I knew I had to do _____ because . . ."

I think one reason I like using this strategy is because it promotes peer feedback and lends itself to students explaining their thinking. I often run into the problem of students just giving each other the answer or just one-word answers. They need a lot more modeling and practice with having academic conversations. This is a shift in my thinking and planning because I need to allow for more time for students to be able to practice these skills and strategies. I do think it's time well spent, though!

One way to model sharing of thinking is to work through a problem for the whole class and say, aloud, what you are thinking as you solve the problem, especially at places where students might get stuck or have a variety of ways to think about the problem. Your model should not only include the steps you are following to solve the problem but also the decision-making process you are going through as you solve the problem and your reasons for any of those decisions. When you do this modeling, let your students know that

- talking through their thinking is something that you value and expect from them when they are asked to explain their thinking, whether to you, another student, or the whole class; and
- they too should share not only their steps but also the decisions they were making about solving the problem and why they did so.

Here's an example of modeling your reasoning and approach to solving a mathematics problem in this way:

Problem:

The cost for taking 4 children to a zoo is $10.00. If Alison's scout troop wants to take all 18 children to the zoo, how much will it cost?
Find two ways to solve this problem.

Teacher:

I'll start with what I know: it costs $10 to get 4 children into the zoo. If I can figure out what it costs for 1 child, I can find the cost for 18 by multiplying. So I want the cost for one child. I'll divide $10 by 4 because dividing splits the $10 evenly among the 4 children. So it costs $2.50 for each child. There are 18 children *in the troop, so I'll multiply $2.50 by 18 to find the cost. (Teacher does the multiplication without comment.) That's $45.*

What's another way to do this? Well, since I have even numbers, I know that it feels easy to me to double them or to cut them in half. So maybe I can use what I know about doubling or halving as another approach to building the cost for 18 children. If I know that 4 children cost $10, then 8 children would cost $20, because I can double both amounts and keep the same relationship between the number of children and the cost of their tickets. I can double that again to know that 16 children would cost $40. But I have 18 children in my troop, so it will be more than $40. How do I figure out the cost of 2 more children? (Teacher pauses.) Well, let me review the problem again with that in mind. Can I use a combination of these doubles of 2 (like 4, 8, 16) to make 18? I can make 18 with 16 + 2, and I've already figured out the cost for 16. How can I find the cost for just 2 children? I notice that 2 children is half of 4 children, and I know the cost for 4 children ($10), so the cost for 2 children must be $5. So adding the cost for 2 children to the cost for 16 gives $40 + $5, which is $45.

The teacher might conclude this modeling with a discussion or reflection that summarizes the important things he or she included, so students will know what they also should include: How to carry out the process is important, but without the explanation of the choices, the process might have been mysterious to someone who used a different approach, so explaining those decisions is important, too.

> **Recommendation 2: Set a clear expectation that everyone needs to participate and contribute, and provide a variety of opportunities for students to contribute.**

This recommendation supports Norm 2 (everyone needs to contribute) and Norm 3 (everyone needs time to think and process). The heart of both of these norms is that everyone matters, and so all students should do their part to help ensure that everyone has a chance to learn.

Make it explicit to your students that every voice matters. Getting a variety of perspectives helps both you, the teacher, and your students to judge how well the class is meeting the success criteria; it also helps an individual student self-assess whether he or she is meeting the success criteria, even if he or she is not the one responding to a particular question or task. Because of that, no one gets to be invisible—overlooked or not required to participate in the class's work; at the same time, no one should dominate a discussion, because doing so prevents other students from contributing.

You can use any of various techniques for distributing the responsibility for adding to class discussions, such as putting names of students on wooden sticks that you draw from randomly when calling on students or using a random generator app on a smartphone or tablet. Knowing that particularly in middle school there can be a sense of safety in numbers, some teachers ease students into sharing their thinking by initially having a pair or a small group work together then decide what the group will present as a whole. Other teachers have taken what could be an intimidating situation for some students and put a fun spin on it (such as in the

Teachers' Voices anecdote below), putting students at ease while still conveying clearly that this kind of risk taking is expected and encouraged.

However you decide to communicate the expectation of participation, getting this expectation across will be an ongoing process. Some students will need more gentle encouragement than others; while it's helpful to be mindful of your students' personalities as you decide how hard to push, it's still important to be sure all students are encouraged to participate, including taking risks from time to time.

TEACHERS' VOICES

In my classroom, I have an established routine that helps to build the social norm for risk taking. At the start of the year, each class is given an empty jar. The collective goal for the class is to fill the jar with marbles.

Students earn marbles for the class in a wide variety of ways: They give innovative answers; they take risks in class; they respond positively to what a peer has contributed; they make a mistake publicly while describing their thinking and mathematical processes; they help another student explain his or her thinking in a way that the teacher can understand it; they ask clarifying questions; they publicly discuss sticky parts of a problem.

Each time I add a marble, I clearly state what the student (or group of students) did to earn the marble. For example I might say, "Ryan, go put two marbles in the Period 2 jar. I really appreciate the fact that you let your classmates give you feedback about the mistake you made in Problem 6 when you put the problem on the board. You earned one marble by volunteering to put your thinking on the board. You earned the second marble by graciously listening to your classmates as they provided you with feedback."

When a jar is full of marbles, the class gets a math party.

While you might explicitly tell students they have a responsibility to contribute, if there are limited opportunities, some students will be unable to meet this responsibility. Not all contributions need to be oral or need to be done in whole-class settings. For example, a student can agree to have his or her work posted for others to learn from or can contribute to his or her small group's work even if someone else will present it or talk about it. Providing a variety of opportunities increases the chances that students will step up with their contributions; students who are reluctant to speak in class may surprise you when they are able to shine in other ways.

When you do pose a question to students who may require some time to think, providing some wait time to allow them to think and formulate an answer can increase the number of students who then have something to say. It's important to enforce the think time and to manage those students who are eager to chime in right away with a response. As part of the social environment, students need to understand why it's important to give everyone a chance to process the question and arrive at their own answer—and what the class can gain by doing so (for example, a variety of students sharing different strategies could provide different insights useful for everyone, even those who are able to find an answer quickly).

For students who need more time to collect their thoughts or formulate an answer, seeing numerous hands up can lead to these students giving up on even trying to formulate an answer, being certain that if they just hang back and remain quiet, someone else will answer the question. To encourage more participation with think time, there are a variety of things you can do:

- Have a no-hands-up rule. Hands can be raised to ask a question but not to indicate readiness or willingness to answer one. You can then use different ways to choose which student to call on to answer the question. For example,

  - when students are ready to respond to a question from you, instead of raising their hands, they show a thumbs-up discretely in front of their chest so that students who are not ready don't see all the hands up, or
  - write students' names on wooden sticks or index cards that you carry with you in a can or a box. After you ask your question, enforce silent thinking time for some time and then randomly select a student to call on by pulling his or her name from the can or box. When doing this, the student called on may not be able to give an answer; you might allow the student to pass or ask a peer for help. The student's name can either go back in the can or box or you might ask the student to paraphrase or explain an answer given by another student.

- Have a countdown timer visible for students, and do not accept responses before the countdown timer has finished.

In addition to helping enforce some think time, these techniques encourage all students to be ready and responsible for providing an answer, since anyone may be called on. Conveying the expectation that you want all students to think and be prepared to respond—and then putting some practices into place that make that possible for students—sends a strong message that students are responsible for being engaged and participating.

Note that waiting can be difficult for some students who arrive at an answer quickly. You might suggest that these students challenge themselves by looking for a second or less obvious way to arrive at the answer or have them record notes about their answer while they're waiting.

Intellectual Curiosity

Our final norm is related to intellectual curiosity:

Sample Norm 4: Curiosity about each other's thinking helps the collective learning of the whole class by providing different ideas. Everyone needs opportunities to provide his or her thinking for others to consider and to hear someone else's thinking, and everyone has a responsibility to listen to others' thinking and give it serious consideration.

Curiosity supports formative assessment in two ways. The first way is that it allows easier and more natural gathering of evidence as students grow more comfortable with your reasons for asking about their thinking. As a result, their responses become less guarded and more forthcoming.

The second way is that, as students express curiosity with each other and listen to each other's thinking, they more easily begin to see their peers as resources. This does not come naturally for many students! Although genuine curiosity is not something that you can insist upon, you can set up an environment for which intellectual curiosity is a natural consequence. You can model this curiosity through your own curiosity for your students' thinking, and there are a variety of ways that you can convey this to students—through body language, comments, wait time, responses to both correct and incorrect comments from students, and questions that you ask.

> **Recommendation 3: Inquire about your students' thinking both when they're correct as well as when they're incorrect.**

This recommendation may seem familiar, as we have a similar recommendation in Chapter 3, "Gathering, Interpreting, and Acting on Evidence," but it also helps you establish Norm 4. You can model the kind of intellectual curiosity you want students to have for each other in the kinds of follow-up questions you pose both during whole-class discussions and in individual conversations you have with small groups or individual students.

Students are faced with many questions during their school day that serve many different purposes, so making sure to include questions that convey a curiosity about their thinking both helps students gain a sense of trust that you really want to know and models for them these questions that convey curiosity about another's thinking. Consider the difference between these two sets of questions:

- If the teacher was focusing on understanding a student's line of reasoning, he or she might point to a table written on the student's paper and ask, "What do these numbers mean?" Or, "How was the table helping you think about the problem?"
- If the teacher was focusing on helping a student follow a process introduced in class, he or she might point to a table written on the student's paper and say, "Look at this entry here. What should that be instead?" Follow-up questions might include one or more of the following: "Now, what's the difference in the y values?" "Next, you look at the y-intercept; what's the y-intercept in the problem?" "Where do you write those in the equation?"

The first set of questions is an *inquiry* about what the student is thinking and how the student is making sense of the problem; this is a necessary part of gathering and interpreting evidence and also sends the message that the teacher is interested in knowing the student's thinking. The second set of questions, on the other hand, *provides instruction* by guiding the student through a particular line of reasoning. These sequences of questions are motivated by a different focus, both of which are necessary at different times during instruction.

If students only get a steady diet of the questions guiding them through a procedure, they are unlikely to be forthcoming with their own thinking, unless they are certain it is already correct. Over time, with an understanding that it's safe to share their thinking and being given time to do that thinking, most students will start to provide you with more and more information about what's going on in their heads. Not only does this give you valuable information about how your students are making sense of the mathematics, especially in relation to the concept-oriented success criteria, but it also serves the purpose of modeling for students how they can be curious about each other's thinking, and that they should be curious about each other's thinking.

Posing follow-up questions after students have made correct responses can be particularly powerful at modeling this curiosity. Sometimes by doing so, you will uncover misconceptions or confusion that may be masked by a correct answer. For example, a student could correctly say that $\frac{4}{5} < \frac{6}{7}$ but have incorrect reasoning that $\frac{4}{5} < \frac{6}{7}$ because $4 \times 5 = 20$, $6 \times 7 = 42$, and 20 is less than 42. Conversely, sometimes you will find that a student does indeed have a thorough and complete understanding of a problem, and you'll gain information about how the student thinks about the problem.

Frequently asking about student thinking sends the message to students that their thinking is important, that you're interested in understanding it, and that this kind of inquiry is part of what one does in math class to learn. You gain more information about how your students are making sense of the content, and they gain a better understanding of what information to give you and a greater sense of trust that it's safe to share their thinking. In turn, this provides students with better insight into what they should be saying, listening for, and asking about, when discussing mathematics with their peers. Whether or not your students are truly curious about each other's thinking (some will be; others will not), as they see this modeled repeatedly during math class, they may start to accept that a high level of inquiry is an important part of collaborating mathematically.

This brings us to our next recommendation, which also helps establish Norm 4:

> **Recommendation 4: Be explicit about the purpose of collaborating.**

As you probably noticed, many of the techniques and strategies we've suggested as part of formative assessment practices provide opportunities for students to collaborate on doing mathematics—for example, working on problems in pairs or small groups; talking with a partner, a small group, or the whole class; and publicly sharing examples of work. As all teachers of middle school students know, social interactions can trump all other priorities and the classroom collaborations of pairs or small groups can sometimes be less than productive. The trick is to help students distinguish between socializing and collaborating, and to do that, you need to talk explicitly to your students about why it's important to collaborate in mathematics class and what that collaboration actually looks like.

Many students will be quick to understand that collaboration means talking to someone else about how you are thinking, but what many students overlook is the flip side to that: the importance of listening to the thinking of others. Make clear to your students that the reason for collaborating in doing mathematics is to gain a better understanding of the topic by understanding different people's ways of approaching and working through a problem, and that this involves listening as well as talking.

Students readily understand the idea of telling someone else how they solved a problem, but they may need some help learning how to listen to what someone else has to say. This is especially true when trying to determine how a peer's comments relate to the learning intention and success criteria. For example, you might provide note-taking templates on which students can record one or two key points about what someone else says. During whole-class discussion, you might provide structure on how to listen and respond. For example, you might want to provide a guiding question, prompting students to ask, *What did you try? What did you think about?* or *How does this show progress toward meeting the success criteria?* Figure 7.3 shows one way that a group of teachers decided to help students understand how to participate in a math discussion as the listener. The figure was provided on a reference card that students could refer to at any time. However, students need opportunities to practice listening and responding with these questions, to help them learn to use the questions appropriately.

Figure 7.3   Participating as a Listener

Any student who is sitting and listening to ideas being shared needs to be ready to be called on to respond in one of the following four ways:

Ask for clarification of what is being said:

> *What did you mean by _____?*

Build on the idea:

> *I agree, because _____.*

Share an alternate approach:

> *I thought about this differently, because ___.*

Test the idea:

> *I'm not sure. Let's try (example).*

 Classroom Material: Go to Resources.Corwin.com/
CreightonMathFormativeAssessment to access this resource.

 Student Summary Card for Participating as a Listener. This printable file is a large-sized card that provides a summary for students for this listening strategy.

Putting Them All Together

Table 7.1 collects the examples of social and cultural norms that we have suggested to support a formative assessment classroom, along with some recommendations to help you establish this part of your classroom culture. When students are given time to think and are faced with the expectation that they will engage with a problem or discussion question, and when they have a sense of intellectual safety to share their thinking particularly when it's imperfect and still sometimes a bit jumbled, it becomes easier to take the risk of making that thinking public. That sense of safety is reinforced when the presiding assumption is that we are all interested in and curious about what each other has to contribute to our collective learning. Students are given both an invitation to and an expectation to collaborate and to contribute to each other's learning.

Table 7.1   Norms and Recommendations for the Social and Cultural Environment

Norms for the Social and Cultural Environment of a Formative Assessment Classroom	Corresponding Recommendations
Sample Norm 1: The classroom is a safe place for thinking and learning. Everyone has the opportunity to discuss their own and others' thinking with respect for the ideas and for the people offering them. All students also have the responsibility to be respectful of others and their ideas.	**Recommendation 1:** Be explicit about what it means to share your thinking and about why it's important.
Sample Norm 2: Everyone in the class needs to participate. Everyone needs the opportunity to add something to whole-class work to contribute to the learning of others, and everyone has the responsibility to look for times in which they can contribute productively.	**Recommendation 1:** Be explicit about what it means to share your thinking and about why it's important. **Recommendation 2:** Set a clear expectation that everyone needs to participate and contribute, and provide a variety of opportunities for students to contribute.
Sample Norm 3: Everyone in class should have time to think and process information. Everyone also has the responsibility to allow more time to those who need it.	**Recommendation 2:** Set a clear expectation that everyone needs to participate and contribute, and provide a variety of opportunities for students to contribute.
Sample Norm 4: Curiosity about each other's thinking helps the collective learning of the whole class by providing different ideas. Everyone needs opportunities to provide his or her thinking for others to consider and to hear someone else's thinking, and everyone has a responsibility to listen to others' thinking and give it serious consideration.	**Recommendation 3:** Inquire about your students' thinking both when they're correct as well as when they're incorrect. **Recommendation 4:** Be explicit about the purpose of collaborating.

■ **THE INSTRUCTIONAL ENVIRONMENT: FRAMING INSTRUCTION TO ENCOURAGE AND MAKE VISIBLE STUDENTS' THINKING AND TO OPTIMIZE LEARNING**

When you think about your classroom environment, you may tend to first think about either the social and cultural environment, as we've just discussed, or perhaps also the physical environment. However, the instructional environment is an important consideration for any formative assessment classroom. A formative assessment classroom's instructional environment is designed to encourage and make visible students' thinking with respect to the learning intention and success criteria and to optimize learning by keeping as much of the intellectual work of learning as possible in the hands of students. This means paying attention to two important elements of instruction: (1) the math tasks and (2) students' responsibility for doing the learning. We provide recommendations for both elements in the following sections.

Math Tasks

One goal of implementing formative assessment practices is to gather information about student learning to inform instruction while that instruction is still underway, so that the students can meet the success criteria. As you've seen throughout this book, success criteria should be aimed at different cognitive levels—for example, some may focus on procedures and procedural knowledge, but others should be more conceptual. The math tasks your students engage in are an important foundation for your gathering of information, and so they should also invite higher-order thinking and help make that thinking visible.

For example, consider the difference between these two problems:

Problem 1: Find a fraction equivalent to $\frac{3}{5}$.

Problem 2: If you choose any two equivalent fractions, Fraction 1 and Fraction 2, and double them, which of the following will be true? Explain why you think so.

a. *The doubled Fraction 1 will be greater than the doubled Fraction 2.*
b. *The double fractions will still be equivalent.*
c. *The doubled Fraction 2 will be greater than the doubled Fraction 1.*
d. *It depends on what the fractions are.*

In Problem 1, a student needs to correctly find an equivalent fraction. In Problem 2, a student needs to analyze a mathematical relationship, make a conjecture about the relationship, and then justify that conjecture. Both kinds of problems have their place, and they might both be needed to elicit evidence of your success criteria, but only one is designed to ensure higher-order thinking.

In the social environment section, we presented Recommendation 3 to help establish a norm around curiosity about others' thinking:

> **Recommendation 3: Inquire about your students' thinking both when they're correct as well as when they're incorrect.**

This recommendation is our first recommendation for the intellectual environment, as well. Implementing this recommendation requires you to embed tasks and questions in your instruction that invite students' thinking—otherwise, there would be nothing to inquire about!

Consider the balance of higher- and lower-demand tasks in the tasks you provide for your students. We recognize that curriculum (which may be prescribed and not at teachers' discretion) frequently drives the instructional activities, and some curricula are better than others at providing rich math tasks that target higher-level thinking. If your materials provide fewer such tasks, you can enrich the existing tasks or discussion by including a few questions that ask the student to

- show with pictures, words, or symbols why something is true;
- explain how a procedure relates to something else they have learned; or
- use various representations to justify a solution method.

If you know of any potential student misconceptions or areas of confusion, you can seed discussion with a question or task that asks students to agree or disagree with a common but faulty line of reasoning and to explain their thinking. For example, questions such as these can spark interesting discussion:

- *Matt says that 0.34 is larger than 0.2, because 34 is larger than 2. Do you agree or disagree? Why?*

- *When Janelle compares $\frac{1}{2}$ and $\frac{3}{5}$, she says that $\frac{3}{5}$ is larger because 3×5 (which is 15) is larger than 1×2 (which is 2). What could you say to Janelle to help her see this method is incorrect?*

- *Maya drew two rectangles that she says are similar, shown here:*

Rectangle A

Rectangle B

- *One of her rectangles is 1" by 2". To find the larger rectangle, she added 1" to the width and to the height so her new larger rectangle is 2" by 3". Do you agree or disagree with Maya's method? Why?*

Not all tasks you provide can, or should, be at the higher levels of cognitive demand. However, when your instructional environment includes a mix of math tasks with varying levels of cognitive demand, you will bring out your students' thinking more easily and elicit evidence of both your procedural and your conceptual success criteria.

Students' Responsibility for Doing the Learning

To further support formative assessment practices, the instructional environment should also be designed to optimize learning by putting the majority of the intellectual work in the hands of the students. This occurs across all the critical aspects of formative assessment. It involves helping students articulate the learning intention and success criteria in their own words so they fully understand them. It involves asking students to do the explaining and discussing that provides evidence of their thinking and then having them learn to use the success criteria to try to gauge the success of their work. And it involves helping students learn how to respond to feedback, and to eventually provide it to others as well, using the success criteria as guidelines for that feedback.

> **Recommendation 5: Strive to keep the intellectual work in the hands of your students.**

Think back on your last few lessons, and consider how much talking you did, versus how much talking your students did; if you are doing most of the talking in every lesson, chances are good that you're also the one doing most of the intellectual work in the lessons. As Dylan Wiliam (2011) wrote,

> *[e]very action that a teacher takes, provided it is intended to result in student learning, is teaching, but the teacher cannot do the learning for the learner; teaching is all the teacher can do. . . . I often say to teachers, "If your students are going home at the end of the day less tired than you are, the division of labor in your classroom requires some attention." (p. 49)*

One of the best ways to optimize your students' learning is to make instructional choices that require *students* to do the thinking, explaining, summarizing, and evaluating whenever possible and sensible, given the lesson. Some different ways you might do this include the following:

- Whenever possible, let students do the explaining. This can be done effectively during a lesson or as a summary at the end of a lesson. It serves the dual purpose of helping students articulate and solidify their own learning and giving you one more piece of information about how your students are making sense of the material.

- Structure math tasks and corresponding discussions that invite students to make the connections between mathematical ideas and to draw conclusions that are the punchlines of a lesson. Sometimes it's necessary to explicitly tell students the key idea, but if they can put it together for themselves and articulate what they understand, they will deepen their learning.
- If you find yourself doing most of the talking in a lesson for more than 10 minutes, with students at most providing brief or one-word responses, pause and take stock of where students' understanding is (Rowe, 1986). Some students can get easily overwhelmed by lots of words and may tune out, so it's important to see what sense they're making of the information you've been giving them. Although there is never enough time in a class period to do all the things you hope, allowing time to pause, check in and gather evidence, and take stock of where the class is can be an invaluable bit of evidence for you. But even more important, *it's valuable for your students*. The opportunity to summarize midstream is an important part of solidifying their learning, so let them do some of the summarizing and restating any important ideas that have come up in the lesson so far.

The References section includes some articles and books, such as "Never Say Anything a Kid Can Say" (Reinhart, 2000), that may provide more ideas for you.

THE PHYSICAL ENVIRONMENT: KEEPING RESOURCES AVAILABLE ■

The way in which you and your students interact physically with your classroom also plays a major role in determining how learning occurs for students. This is something that all teachers understand implicitly, and they make careful choices when setting up their classrooms. Thinking about the physical environment in your classroom may bring to mind questions about arrangement of space and organization of materials. For example, table groups of four students facing each other in pairs provides easy ways for students to work in groups but requires students to turn to see the front of the room. Other arrangements may be more helpful for keeping student attention on the board and the teacher but make peer conversation a bit more difficult. Materials should be accessible when needed but out of the way when they're not. While these remain important for a formative assessment classroom, we focus here on keeping available for your students the resources that are specific to supporting formative assessment.

> Recommendation 6: Arrange your classroom to allow students easy access to resources, including learning materials, evidence-gathering materials, self-assessment materials, and their peers.

In a classroom using formative assessment practices, particular resources are always available to the students and to the teacher. The most critical are the learning intention and success criteria. Easy access to this information helps students as they begin to independently gauge their own learning. It also allows students to evaluate their peers' work and serves as a basis for offering feedback. Having the information easily accessible yourself allows you to relate the learning throughout a lesson back to the central learning intention, clearly and (depending how you are supplying it) in a visual way, and to provide both individual and whole-class feedback.

As we noted in Chapter 2, "Using Mathematics Learning Intentions and Success Criteria," there are different ways in which you can ensure your students have access to the learning intention and success criteria. The most common way is to post them in a large, readable format, preferably in a place that you and your students will look at often. As you refer to the learning intention and success criteria throughout a lesson, standing next to or pointing to them, students will also become accustomed to referring to them in this predictable location.

The solution of how to physically manage the display of this information can feel tricky, particularly for teachers who have multiple math classes each focusing on different learning intentions coming to them during the day or who teach multiple subjects to the same group of students. Teachers we have worked with have tried a variety of different strategies for keeping this information separate for different classes, including

- putting the learning intention and success criteria at the top of relevant handouts for the lesson,
- having students copy the learning intention and success criteria into their notes,
- using large books of chart paper bound with rings or using easels that allow teachers to flip back and forth between different charts with different learning intentions and success criteria for different classes,
- printing the learning intention and success criteria onto labels that can be peeled off and attached to students' handouts or work, and
- writing them on large whiteboards dedicated to this purpose.

Figure 7.4 This Teacher Devotes a Section of the Bulletin Board to the Learning Intention for Each Grade She Teaches.

Figure 7.5   This Teacher Uses an Easel.

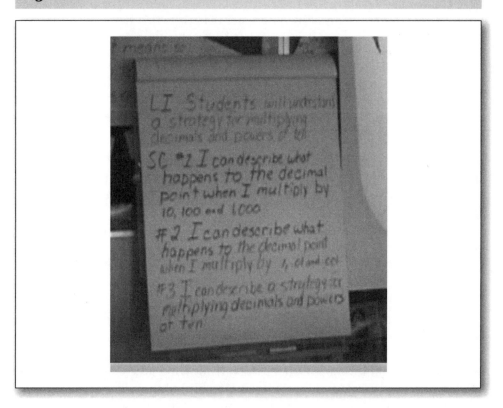

Figure 7.6   This Teacher Uses Sentence Strips in Pockets and Color-Codes Each Class for Easy Switching.

Learning Resource: Go to Resources.Corwin.com/ CreightonMathFormativeAssessment to access this resource.

Images of Posted Learning Intentions and Success Criteria. This file includes images of the various ways teachers have posted learning intentions and success criteria for reference and use during a lesson.

Many things are possible; find what works for you! Whatever technique you use, the important goal is to have the learning intentions and success criteria available for reference by students and by you throughout the lesson.

Let's look at the other kinds of resources mentioned in the recommendation.

Learning Materials

Many teachers set up routines in which students get and hand out the learning resources needed for an activity, such as manipulatives, calculators, graph paper, or various templates. Typically the teacher will guide them on which items to get for the lesson. However, allowing access as needed gives students some control (and therefore ownership) of their learning.

For example, one class was working on integer operations, using a colored chip model. (In this model, one color represents a positive unit, and another color represents a negative unit.) When the teacher began to transition them away from the model to abstracting the rules they discovered using the model, some students found it helpful to pull out the chips to check their work. In another class working on using nets of prisms to understand surface area formulas, one student decided to copy a given net onto available graph paper so she could create the prism. In both examples, not only did the students find it helpful to have a manipulative to help them visualize what was otherwise a fairly abstract problem, but the teacher also saw their choice of resources as evidence that the students were partially meeting the success criteria, needing a little extra help to be confident in the work they did without use of a concrete model.

It's also useful to create a dedicated space to display relevant student work, pictures, diagrams, or other items that can be used as references to move students' learning forward on a particular mathematics concept, particularly if students have opportunities to walk over to this space and compare their work to the examples provided. If a class has discussed examples of work that meets the success criteria, then keeping those examples publicly available provides a valuable reference point for students as they evaluate their work while it's underway. In addition, teachers can use the work as touchstones when providing feedback, encouraging a student to refer to certain pieces of work that might provide a model or example that the student can use. Some teachers keep files of exemplars (and faulty work for targeted discussions) taken from past classes' work, to be used for these purposes.

Evidence-Gathering Materials

Easy access to evidence-gathering materials—such as whiteboards, response cards, and traffic light cards or cups—allows you to elicit evidence more freely throughout the lesson without spending time distributing these materials. While you might have some planned opportunities, having these materials always available widens your options when you have a spontaneous need to elicit evidence. In addition, using these items regularly helps students understand that they are responsible for providing information about their learning and, perhaps more important, it underscores the message to students that they share some responsibility for monitoring their own learning.

Self-Assessment Materials

A classroom using formative assessment practice also makes available materials that assist students in self-assessment. This communicates to students that self-assessment is an everyday part of what they do in math class. As they learn to use these materials and see others using them, self-assessment becomes more an anticipated part of their classroom culture.

For example, in one classroom, the teacher designated a part of her bulletin board as a place where students could indicate their own assessment of their current understanding of the topic by moving construction paper owls with their names on them into the red, yellow, or green zone (Figure 7.7). Students were encouraged to independently change the position of their owls as their understanding evolved.

In another classroom, the teacher asks students to place a colored cube in a bin as they're leaving the room to indicate their feeling about their level of understanding. A green cube indicates feeling like they have solid

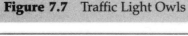

Figure 7.7 Traffic Light Owls

understanding, yellow indicates a partial understanding or still feeling unsure of their understanding, red indicates no or almost no understanding. The teacher creates sections in the bin to easily be able to evaluate the general frequency of different responses (Figure 7.8). Note that, unlike the Traffic Light Owls, this cube method allows some measure of anonymity, since the students' names are not on the cubes.

Figure 7.8   Traffic Light Bin

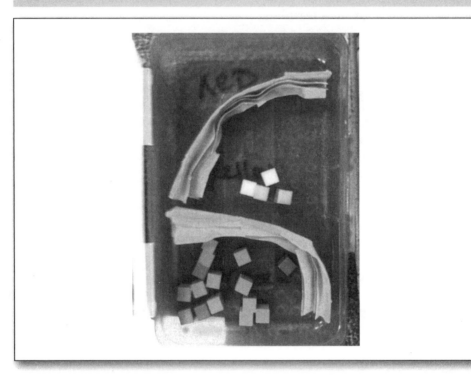

Peers as Resources

A crucial part of implementing formative assessment practices is helping students make their thinking transparent to each other, either by displaying it or discussing it, then evaluating that thinking in relation to the learning intention and providing feedback accordingly. So promoting interchange between students—including but not limited to working in pairs or small groups or easily coming together as a whole class—can support the use of peers as a resource.

At times you will want students to work alone, in pairs or small groups, or as a whole class. If you have movable seating arrangements, consider how you might have students work with you to easily make these arrangements possible. For example, one teacher typically has her students' desks arranged in a U for whole class interaction. Students can easily turn to a neighbor for pair work or discussion, and two sets of pairs can move their desks for small groups. When students are to work alone, the teacher calls for "Office Space," and the students move their desks forward or backward to give themselves more space. They learn early in the year whether they are to move forward or backward.

If you don't have movable seating arrangements, or simply don't have room to vary your seating arrangements, consider other kinds of meeting space in your room. For example, one teacher designated a meeting spot she calls the "Meeting of the Minds," where students can stand, with or without the teacher, to share ideas about something they're working on. Another teacher uses a small rug to designate a space on the floor where students can sit when they need help from her or from each other. This strategic use of space can have an added benefit of allowing students who might be reluctant to speak out in a formal whole-class setting to talk more naturally in a group. In her work, Claire Lee, a British mathematics and formative assessment researcher, noted that "even pupils who in other situations may have been reluctant to answer joined in the conversations when they were surrounded by others struggling to understand a mathematical concept" (2006, p. 23).

Reference Resource: Go to Resources.Corwin.com/ CreightonMathFormativeAssessment to access this resource.

Chapter 7 Summary Cards. This printable file provides Summary Cards showing a summary of the elements of classroom environment to consider.

CONCLUSION ■

In this chapter, we've focused on those elements of the classroom environment that have the most direct impact on implementing formative assessment practices effectively. The social and cultural environment plays a key role through the norms for behavior and personal interactions that support students' gradual movement toward becoming more self-regulating. The instructional environment plays a key role in establishing what mathematics opportunities students will have and what role the teacher and student will play when given those opportunities. The physical environment plays a key role in the ways in which it can make resources (including materials, information, and peers) available to students.

We've now discussed all six aspects of formative assessment:

- The four critical aspects

    o Learning Intentions and Success Criteria
    o Eliciting and Interpreting Evidence
    o Formative Feedback
    o Student Ownership and Involvement

- The two supporting aspects

    o Learning Progressions
    o Classroom Environment

In the final chapter, we provide some principles and final recommendations for implementation that can help you as you begin to weave all these elements together.

Recommendations in This Chapter

1. Be explicit about what it means to share your thinking and about why it's important.

2. Set a clear expectation that everyone needs to participate and contribute, and provide a variety of opportunities for students to contribute.

3. Inquire about your students' thinking both when they're correct as well as when they're incorrect.

4. Be explicit about the purpose of collaborating.

5. Strive to keep the intellectual work in the hands of your students.

6. Arrange your classroom to allow students easy access to resources, including learning materials, evidence-gathering materials, self-assessment materials, and their peers.

Reference Resources: Go to Resources.Corwin.com/ CreightonMathFormativeAssessment to access these resources.

Chapter 7 Summary Cards. This printable file provdes Summary Cards that include recommendations from this chapter.

Formative Assessment Recommendations. This printable file provdes a summary of all the recommendations in this book.

■ RESOURCES

The following sections present some resources that can help you implement the recommendations. Each of these resources is referenced in earlier sections of the chapter, but here we provide a consolidated list. All resources can be found at Resources.Corwin.com/CreightonMath FormativeAssessment.

Learning Resources

These resources support your learning about elements of the classroom environment related to formative assessment practices:

- **Images of Posted Learning Intentions and Success Criteria** shows various ways teachers have posted learning intentions and success criteria for reference and use during a lesson.

In addition, videos are available to support your learning about formative assessment practices related to the classroom environment.

Reference Resources

These resources summarize ideas about elements of the classroom environment related to formative assessment practices:

- **Chapter 7 Summary Cards** provides index-sized Summary Cards for the following:
  - **Considerations for Environment to Support Formative Assessment** includes brief descriptions of each of the elements described in this chapter.
  - **Recommendations for Classroom Environment** includes all the recommendations from this chapter.
- **Formative Assessment Cycle** is a PDF of the color-coded, full version of the Formative Assessment Cycle.
- **Formative Assessment Cycle Video** is an audiovisual tour of the completed Formative Assessment Cycle.
- **Formative Assessment Recommendations** is a printable PDF that includes all the recommendations from this book.

Classroom Materials

This resource can be copied or re-created to be used in your classroom:

- **Student Summary Card for Participating as a Listener** is a large-size card that provides a summary for *students* for the listening strategy in the chapter.

8

Moving Toward Implementation

We've now addressed all the critical aspects of formative assessment: learning intentions and success criteria, eliciting and interpreting evidence, using formative feedback, and building student ownership and involvement. We've also addressed the two crucial supporting aspects: understanding and working from learning progressions and developing a classroom environment conducive to implementing formative assessment practices.

We completed the Formative Assessment Cycle in Chapter 7, but let's take this opportunity to step back and look at it again in Figure 8.1. Whether you're now beginning to try out some of the pieces, or you're already familiar with them separately, weaving them together requires some focus on formative assessment as a *process*, and the cycle diagram illustrates how the many pieces—all dependent on other pieces—fit together.

Reference Resources: Go to Resources.Corwin.com/
CreightonMathFormativeAssessment to access these resources.

Formative Assessment Cycle. This is a color version of the completed Formative Assessment Cycle, printable on legal (8.5" by 14") paper.

Formative Assessment Cycle Video. This is an audiovisual tour of the Formative Assessment Cycle that we have built throughout this book.

Figure 8.1 The Formative Assessment Cycle

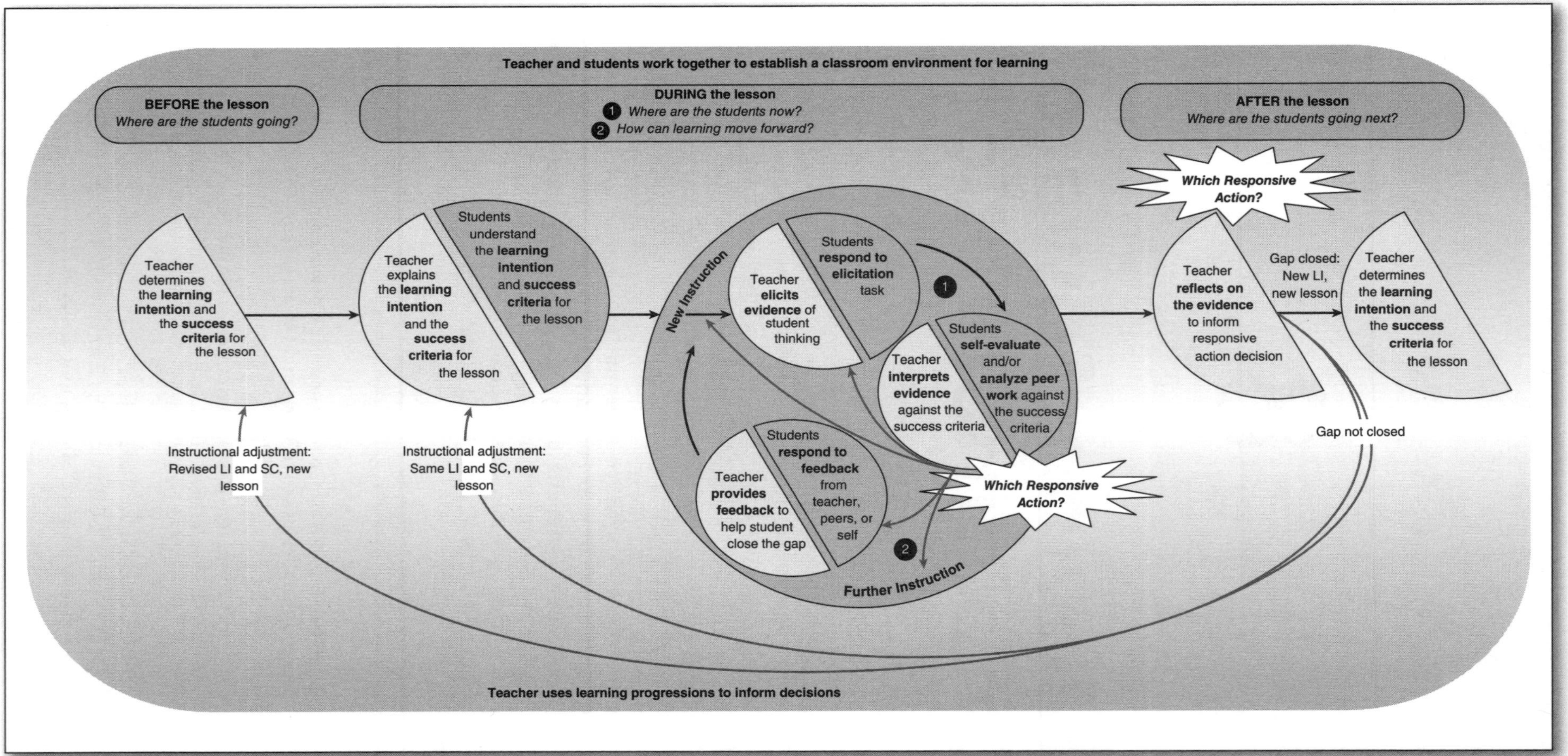

Teacher and students work together to establish a classroom environment for learning

BEFORE the lesson
Where are the students going?

DURING the lesson
1 Where are the students now?
2 How can learning move forward?

AFTER the lesson
Where are the students going next?

Teacher determines the learning intention and the success criteria for the lesson

Teacher explains the learning intention and the success criteria for the lesson

Students understand the learning intention and success criteria for the lesson

Instructional adjustment: Revised LI and SC, new lesson

Instructional adjustment: Same LI and SC, new lesson

New Instruction

Teacher elicits evidence of student thinking

Students respond to elicitation task

Teacher interprets evidence against the success criteria

Students self-evaluate and/or analyze peer work against the success criteria

1

Which Responsive Action?

2

Teacher provides feedback to help student close the gap

Students respond to feedback from teacher, peers, or self

Further Instruction

Which Responsive Action?

Teacher reflects on the evidence to inform responsive action decision

Gap closed: New LI, new lesson

Teacher determines the learning intention and the success criteria for the lesson

Gap not closed

Teacher uses learning progressions to inform decisions

With all the different moving parts that come together to create a vibrant classroom taking full advantage of the power of formative assessment practices, you might find it difficult to figure out where to begin. In this chapter, we give you a closing overview of the information in the first seven chapters, in the form of a list of principles that can guide you as you implement and connect the formative assessment practices. Then we give you some specific recommendations and tools for getting started, or if you've already started, for putting all the pieces together.

IMPLEMENTATION PRINCIPLES FOR FORMATIVE ASSESSMENT ■

As we worked with teachers in our professional development work, we provided some guidelines to help them to set goals. We arrived at a set of general guiding principles, each with a set of indicators to help them assess the extent to which they were successfully implementing formative assessment practices. We list the principles here, and we provide the indicators in Appendix B.

Learning Progressions

A. Be clear about what's mathematically important or significant in your content (the concepts as well as the procedures) and how that content connects with other mathematics and grows both within a unit and across units.

When you have incorporated this principle in your teaching, you think about the progression of concepts across the unit or chapter you will teach, perhaps using an applicable learning progression as a guide.

Learning Intentions and Success Criteria

B. Learning intentions help your students focus on what they should be learning, and success criteria help them gauge the extent to which they are learning it.

Your learning intention makes clear to students what the central focus of learning is for the lesson. The corresponding success criteria focus on tangible indicators of reaching the learning intention, provide enough evidence of the learning described in the learning intention, and are aligned to the learning intention. Your learning intentions focus on developing conceptual understanding of underlying mathematical ideas. Success criteria include both procedural fluency and evidence of higher-order thinking (such as explaining or justifying).

C. You and your students both need access to the learning intention and success criteria throughout the lesson.

You might post them publicly somewhere in the room or put them on the top of a handout. What's important is that they remain accessible throughout the lesson, so both you and students can refer to them at any time. As you connect tasks and feedback to the success criteria, pointing back to them will help students learn that they are there as a reference for them to use.

Eliciting and Interpreting Evidence

D. Students need opportunities and encouragement to articulate the reasoning behind their work.

You provide time for students to discuss or share their thinking. By applying this to correct work as much as to incorrect work, not only can you gauge how well they are meeting the success criteria, but also you communicate to students that articulating their reasoning is an important part of mathematics learning.

E. Students need opportunities to provide evidence of each of the success criteria.

Your lesson plan includes at least one opportunity for students to demonstrate the extent of their success with each criterion. For example, if one of the success criteria asks students to explain the relationship between two mathematical ideas, you ensure that your lesson includes an opportunity to discuss, share, or write that explanation.

Spontaneous opportunities to gather evidence will arise as well, but planning for the evidence you want to collect can focus your lesson planning. For example, if you know some of the common barriers or misconceptions that students are likely to have with a mathematics topic, you build into your lesson a way to find out if your students have those misconceptions.

F. You need variety in how, when, and from whom you gather evidence, to get a full picture of the class's understanding, as a whole.

This variety is important in several different ways:

- You need variety in the set of students from whom you gather evidence, so you sometimes gather evidence from every student in the class and sometimes from selected students. In addition, you may not need to assess each student to the same degree of depth within a given lesson.
- You need variety in when you gather evidence, so you gather evidence at various points during a lesson, not just at the end.
- You need variety in the type of evidence you gather over time, so you sometimes gather evidence of skill mastery, and you sometimes gather evidence of student reasoning or understanding (through means such as explanations or justifications).

G. All gathering and interpretation of evidence needs to be in terms of the success criteria.

In order to manage the amount of evidence you gather and to be able to use it effectively, you prioritize it by focusing on gathering evidence that is specifically related to the success criteria.

You then interpret that evidence against the success criteria to determine an appropriate responsive action. In many cases, with a class full of students, you may be able to group students who have similar needs for various next steps.

Formative Feedback

H. Formative feedback communicates to the student how he or she has met, as well as not yet met, the success criteria, and it provides hints, cues, or models for next steps.

When giving formative feedback to students, you ensure that they understand both what they are doing or understanding correctly, with respect to the success criteria, as well as what they are not. This feedback helps the student understand specifically the extent to which his or her work meets the success criteria. Your feedback also includes specific guidance on next steps in the form of a hint, a cue, or a model so that students can take the next steps themselves

I. Formative feedback should be focused and specific.

Your feedback targets only one or two things at a time; it does not include all at once a complete list of all the things the student needs to address. This ensures that you provide a manageable amount of feedback to which the student can respond.

Your feedback also focuses on specific elements of the student's work, so that the student knows *which parts* of his or her work meets the success criteria and which parts need revision.

J. You need to prepare yourself to provide formative feedback and plan opportunities for students to respond to it.

You prepare yourself to provide formative feedback by making very clear to yourself what it will sound like or look like to meet the success criteria. Rather than simply describing in general terms what students would say, drafting actual responses you would hope to get if a student has ideally and fully met the success criterion provides you with a clear target you can point students to with your feedback. For example, consider a success criterion that says "I can describe the connection between the two patterns." Rather than saying to yourself, "Students should tell me how the two patterns are related," aim for saying, "Ideally, I would like to hear: 'The x values in the number pattern match up with the figure number in the picture pattern, the y values match up with the total number of squares in each picture, and they go up by two in the table because two squares are added to each new consecutive figure.'"

You also take time to ensure that students understand the feedback they receive. In doing so, you communicate to students the importance of taking action on the feedback they've received. You plan opportunities for students to respond to the feedback and hold students accountable for responding to it.

Student Ownership and Involvement

K. Students need to understand what the learning intention and success criteria mean.

You give students an opportunity to think about what new learning intentions and success criteria mean and how meeting the success criteria shows that the learning intention is being attained. You discuss new terminology or share and discuss an example of work that meets the criteria to help illustrate what the criteria mean. It's not the same as giving away the answer; it's creating a common understanding of what a rubric for success includes.

You revisit the success criteria during the lesson to help students focus their work and to solidify understanding of what the criteria mean, how to meet them, and how meeting them connects back to the learning intention. Students may not always be clear at the start of a lesson what the wording of a particular success criterion means, or what meeting that criterion could look like, and revisiting after they have had a chance to get further into the lesson can help them understand these things better.

L. Students need to use the success criteria to guide their self-assessment and self-monitoring.

You plan for and provide opportunities for student self-assessment on a regular basis. Structuring such opportunities includes giving students self-assessment tools (such as templates, handouts, or other resources) as well as providing time and opportunity either during or outside of class.

Initially, you model for students how they can use the success criteria to

a. gain clarity about what they're supposed to be learning and

b. self-assess their work.

Over time, you will model this less as students adopt the habit of checking the success criteria in these situations.

M. Students need to learn how to provide and use evidence of their mathematical reasoning and understanding in their self-assessment, and they need encouragement to do so.

You help students understand how the evidence gathered relates to the success criteria, and you provide opportunities for students to self-assess their work. You might model what it means to share their thinking with the class or model ways to use the evidence to self-assess their work against the success criteria. You help students use evidence of their actual performance to determine their level of confidence.

Over time, your students become comfortable explaining their thinking, both when their solution is correct as well as when it is not; in particular, your students understand that being questioned does not imply that they are incorrect.

N. Students need to understand what feedback is, why it's helpful, and how they should use it.

You devote some class time to helping students learn about formative feedback, practice using it, and understand how such feedback is based on the success criteria.

O. Students need opportunities to provide and receive peer feedback, as a way to learn to provide feedback to themselves about their own work.

Structuring opportunities includes giving students tools for providing and receiving peer feedback (such as templates, handouts, or other resources) as well as providing time and opportunity either during or outside of class to practice providing and receiving the feedback. You are also explicit with students about the goal of eventually doing the same thing for themselves about their own work as part of self-monitoring.

P. Students understand a variety of ways to act on the results of their self-monitoring to take the next steps to move their learning forward.

You can model for students how to make decisions about next steps (refer to a resource, rework the task, request feedback, etc.). You build in opportunities for students to practice making these decisions about their own work and checking them with a peer or the teacher. You then build in opportunities for students to practice following through with next steps to improve their work.

Classroom Environment

Q. To support formative assessment practices, the social/cultural environment of the class promotes intellectual safety and curiosity.

You establish norms and guidelines that set expectations about students' participation, including sharing their thinking and responding to the thinking of others. You provide opportunities for them to participate, including collaborative activities and allowing appropriate think time so all students have the chance to process your questions. You also explore both correct and incorrect thinking.

R. To support formative assessment practices, the instructional environment optimizes learning and encourages and makes visible students' thinking.

You frame tasks and discussions in ways that invite students' thinking, and you keep the intellectual work in the hands of the students as much as possible.

S. To support formative assessment practices, the physical environment allows easy access to various resources.

You have a predictable location in which you make available the learning intentions and success criteria. You create a physical space that maximizes opportunities for interchange between students. You also provide easy access to any evidence-gathering or self-assessment tools for students.

Reference Resources: Go to Resources.Corwin.com/ CreightonMathFormativeAssessment to access these resources.

 Implementation Principles for Formative Assessment. This resource provides a list of the principles described earlier.

 Implementation Indicators for Formative Assessment. This resource includes the full list of the principles and associated indicators (also in Appendix B).

 Correlation of the Principles, Recommendations, and Resources. This resource provides the principles listed above, correlated with the associated recommendations and resources.

■ ## SUSTAINING YOUR EFFORT OVER THE LONG TERM

Given everything we've laid out about all the different practices of formative assessment, you may be wondering "But how do *I* do this? What's the best way to begin? How do I manage all of this?" The following recommendations are intended to provide some overall guidelines, and the accompanying resources comprise a set of tools to help you manage your implementation and sustain your effort over the long term. We talk about some ways to use these resources in the following recommendations.

Within the recommendations, we also share some important lessons we've learned about the implementation of formative assessment from our 5-year professional development project, Formative Assessment in the Mathematics Classroom: Engaging Teachers and Students (FACETS), funded by the National Science Foundation. During the project, we were privileged to work with two cohorts of teachers, each for 2 years. These teachers taught students in Grades 5 through 8, and while all taught mathematics, some were also special education teachers. Each was involved in the professional development that resulted in this book. Yet, while all participated in similar professional development, each teacher followed his or her own path toward implementing some or all of the formative assessment practices outlined in this book. At the end of their professional development work, here is what some of the FACETS teachers said:

> *I can [say that] what worked for me, was taking what I was doing and trying to apply what seemed to make sense to me, at first; doing little bits at a time, sticking with it and sort of letting it set in and grow roots, and being willing to kind of push yourself a little bit to try things.. . . So try and take what works, and meld it to what you do. (Grade 8 teacher)*

> *It's a multifaceted approach to teaching. . . . There are going to be pieces that you yourself can anchor into, that feel familiar . . . usually everybody has something they feel they can anchor to, so anchor to that, and don't worry about the fact that the other things feel really fuzzy. . . . But it doesn't take much to get started; for me, it took courage to keep going with these little itty bitty pieces, knowing that the little itty bitty pieces were making a huge difference even if I didn't have the global picture. And the other pieces snapped into focus for me over the time period I worked with it, very dramatically at very unexpected moments. And every time something has snapped into focus, the whole picture has gotten easier. . . . I feel more relaxed, I feel more capable of teaching math . . . my ability to confidently feel like I know what my students are learning and how to move them forward has shifted. (Grade 6 teacher)*

Their comments here capture an important message about implementing these formative assessment practices: It's important to graft them onto to what you already do and make them your own. In this book, we provide recommendations based on principles of formative assessment, but these recommendations are not recipes. There are no prescribed ways to implement each of these practices; you can incorporate them into your own teaching practice in a way that fits your teaching style. This flexibility presents both a challenge to learning to implement them, particularly early in your work with formative assessment, but it also can result in a powerful and lasting impact for you and for your students. We cannot predict what your own path will be in implementing these practices, but we can share with you some of what we learned from the paths taken by teachers in the FACETS project.

> ### Recommendation 1: Be kind to yourself; don't try to do everything all at once.

As many of the teachers would concur, it can be overwhelming at first to think about trying to incorporate all of the pieces of formative assessment into your instruction. However, many of them agreed that initially focusing on one thing at a time was the way to make it manageable. Here, *focusing on one thing at a time* has a couple of different meanings. The first meaning refers to focusing your learning on one critical aspect at a time, when you're first starting out. There are many moving parts to formative assessment, so give yourself permission to take it slow and work on one at a time. The other meaning refers to one of the important lessons learned that emerged from the professional development: although it's important to help your students learn how to step into their role within formative assessment, it's also important to give yourself permissions to focus on your own learning first.

Lesson Learned: It's Important to Give Yourself Permission to Focus on Your Own Learning First

Throughout this book, we have emphasized the power of using formative assessment practices as a mechanism to support *students* in developing more ownership in and involvement in their mathematics learning. However, before focusing heavily in this area of the students' role in formative assessment, it makes sense to feel comfortable with *your own* role in the various aspects of implementing formative assessment. We encourage you to give yourself permission to focus on your own actions at first; initially, it can be sufficient to just keep in mind the students' eventual role in the process.

In our own experiences working with teachers on formative assessment, we've discovered that many teachers needed to focus on their own actions initially; many of them were busy wondering, "What am *I* supposed to be doing?" and "Am I doing it correctly and effectively?" We characterize this as a time of gaining familiarity with the aspects and figuring out the management of the different formative assessment practices.

As the teachers we worked with became more comfortable with their role, they were more easily able to focus on their students' role in formative assessment. Their questions turned more toward "How can I help my students be more proactive about managing their learning and about communicating their understanding?" We characterize this as a focus on empowering students to take part in formative assessment (see Figure 8.2).

Figure 8.2   Shifting From a Focus on Familiarity and Management to a Focus on Empowering Students

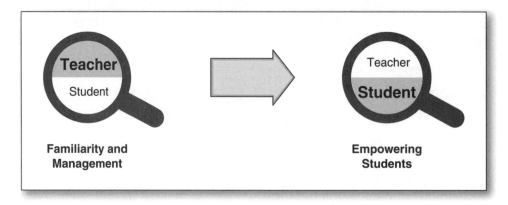

We can also relate this shift to self-regulation (see Figure 8.3):

Figure 8.3   Thinking Like a Self-Regulating Learner

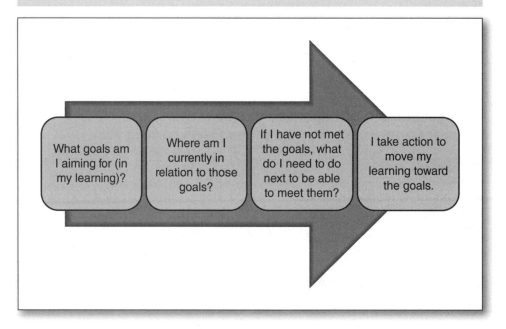

In earlier stages of implementation, your attention may tend to focus more on how *you* answer the various self-regulation questions *about* your students:

- What goals am *I* (the teacher) aiming for in my students' learning?
- Where are they (my students) currently in relation to those goals?
- If they have not met the goals, what do *I* need to do next to help them be able to meet them?

These questions are not terribly different from what you have to think about everyday as a teacher. Because these questions are so familiar, we found that during initial efforts to start integrating formative assessment practices into instruction, the teachers we worked with were often wondering, "What would I do to implement formative assessment differently than what I already do?" If you find yourself wondering the same thing, it may help to know that that's a part of focusing on yourself first to learn what you need to do.

As the teachers we worked with gradually came to understand each of the aspects, and understand their role in relation to each of those aspects, they moved into the Empowering Students part of their implementation. (See Figure 8.2.) At that point, their instructional focus shifted more heavily toward helping students learn to answer the self-regulation questions for themselves:

- How do I help my students know what goals *they* are aiming for their learning?
- How do I help them understand where they are currently in relation to those goals?
- If they have not met the goals, how do I help them understand what to do next to be able to meet the goals?
- How do I help them understand how to take action to meet the goals?

But before you can get there, allow yourself the learning curve you need to become comfortable enacting your role, as the teacher, in formative assessment.

Lesson Learned: This Work Takes Time

You may have noticed that a mention of "taking the time" has come up repeatedly throughout the book. It does indeed take time to learn both about the individual aspects and how to integrate them, and more time to then bring students into the formative assessment equation.

That said, we have no magic number for how long it takes teachers in general to learn this; it depends on what any teacher brings to this work in terms of mathematics content knowledge, instructional style, repertoire of pedagogical techniques, and familiarity with being analytic and reflective about student learning. Many of the building blocks of formative assessment are based in the ordinary things that you do already: sharing learning goals with students, gathering information about what students know, providing feedback, and making instructional decisions about how to proceed based on how students are doing. However, in formative assessment, there are nuanced and important ways to carry out each of these things that may or may not be different from what you already do. Understanding these nuanced practices—why they are important and then how to carry them out—is an important part of your learning to implement formative assessment practices. Our professional development program spanned 2 years, and even then, many teachers expressed a desire for more time to work together because they wanted to continue their momentum with support and collaboration. As with any important instructional initiative, learning to implement formative assessment will take time, but the investment of time and effort will pay off dramatically. So be kind to yourself, and celebrate small steps of progress along the way.

> **Recommendation 2: There is benefit in collaboration! Seek support for this work.**

Maintaining momentum and energy for a new professional learning goal can be hard work, especially when you are working independently. It can be very helpful to find a thinking partner with whom you can do some planning and some sharing of ideas or strategies. The benefits of collaboration were very apparent throughout the professional development work we did.

Lesson Learned: Teachers Benefited From Working Collaboratively

Because collaboration promotes professional learning, we required that participants join our professional development program as school partners or teams. While most of these partnerships and teams were maintained for the duration of the program, some teachers in each cohort lost their in-school collaborators, for various reasons. These teachers told us that they wished they had someone to work with in their building and that working alone was more challenging.

Teachers found it easier to write learning intentions and success criteria when working collaboratively with someone else. Some teachers who had common planning time with colleagues would meet regularly to plan lessons, articulate a learning intention and success criteria for the lesson, discuss possible misconceptions or stumbling blocks for students, and plan how to build in opportunities in the lesson to gather evidence of those potential difficulties.

In addition, we learned that having collaborators also made it more likely that teachers could continue their efforts long enough to start to see some results with their students. As we've mentioned throughout the book, learning to implement formative assessment practices takes time: time to learn what to do in your role, then time to help your students learn how to step into their role. You'll need some patience to continue your work to the point where you can see the payoff for your students. One teacher felt that learning to implement these formative assessment practices felt analogous to learning to play the piano: You needed to be patient enough to get past learning the mechanics with *both* hands before you could begin to sound the way you'd like to sound.

We encourage you to find at least one other person to work with. If you are able to find a colleague with whom to collaborate, you might consider doing any of the following:

- Choose an area of formative assessment in which to focus your learning—possibly one of the aspects but perhaps a smaller part of one of them—and then familiarize yourself with the indicators (Appendix B) for that area. You might decide to work on the same area as your colleague, but valuable collaboration doesn't require this.
- Pick and choose from the indicators related to your focus area. Focusing on no more than two or three at a time is a very reasonable goal. And sometimes, just one is enough.

- Get together to plan lessons, and use these indicators to guide these discussions. Keep doing whatever you've already been doing in your planning, and then look for small ways to start to infuse elements into your lesson that address the indicators that you've chosen to work on.
- If you decide to observe each other and give each other feedback, the indicators you've chosen can serve as a focus for your observations of each other and for the conversation about the lesson after the observation. Consider them your success criteria for your implementation goal.
- Consider having a colleague video a lesson so that together, you can look over the questions you ask and identify how often you ask higher-order questions versus lower-order questions and how these questions align with your lesson's success criteria.

Learning and implementing these practices on your own is still worthwhile. If you are in a situation where you are working on this by yourself, focus on developing a workable plan for yourself in which you can gauge your own progress. (See Recommendation 3, which includes a reference to a self-reflection and planning resource for this purpose.) We encourage you to return often to the recommendations throughout the book and to take them to heart!

Reference Resources: Go to Resources.Corwin.com/ CreightonMathFormativeAssessment to access these resources.

 Formative Assessment Recommendations. This printable file includes a summary of the all recommendations in this book in a printable (letter-sized) format.

 Implementation Indicators for Formative Assessment. This resource includes the full list of the principles and associated indicators (also in Appendix B).

In either of these situations—whether working individually or collaborating with a colleague—continue to explore ways to secure support for yourself. In our experience, teachers in the FACETS project had greater success in their implementation when they had the support of colleagues, coaches, or administrators. You may find it helpful to share information about formative assessment in mathematics with an administrator or colleague who wants to support you and to understand what you're working on. We can recommend any one of a set of articles listed in the Resources section of this chapter.

Reference Resource: Go to Resources.Corwin.com/ CreightonMathFormativeAssessment to access this resource.

 Reference Resource: Books and Articles List. This printable file provides a list of potential resources to share with others.

> **Recommendation 3: Use a self-reflection tool to check in on your progress, set new goals, and adjust your path.**

In our work, a question that we heard very frequently in teachers' early efforts to implement these practices was, "How do I know if I'm doing it?" or "How do I know if I'm being successful?" We heard this as a request for guidelines to determine what being successful looks like—in effect, a list of your own success criteria! The recommendations listed in this chapter provide a set of general guidelines for rolling out this work in your own classroom, but we also know that it can be helpful to have a hands-on tool that can help you narrow down specific next steps for your own progress.

We've created the "Reflecting on Your Implementation" tool, based on several self-assessment activities we did with the teachers in the FACETS project. This tool contains the principles and the indicators with instructions for self-assessment. We encourage you to use the tool to develop a starting plan for your learning and implementation of formative assessment and to then return to it on an ongoing basis (for example, once a month or two or three times a year) to assess your progress and adjust your plan to set new goals. In keeping with the prior recommendation, if you have the opportunity to work with a colleague or group of colleagues, this collaboration can enhance your plan.

 Learning Resource: Go to Resources.Corwin.com/ CreightonMathFormativeAssessment to access this resource.

 Reflecting on Your Implementation. This printable file provides indicators for assessing the extent to which you are incorporating the principles into your instructional practices and considering next steps for gradually moving toward full implementation.

Although we structured the professional development for the teachers to unfold in a certain way, addressing various critical aspects of formative assessment in a particular order, we found that over time teachers' individual interests and professional learning needs began to diverge from one another. This led us to another important lesson learned, which we have mentioned before: Choose a path through implementation that makes sense for you.

Lesson Learned: People Can Take Different Paths in Learning to Implement These Practices

As our professional development work progressed, some of the teachers were particularly intrigued by the idea of delving into formative feedback, while others wanted to explore evidence, or to do more work with learning progressions. The various aspects of formative assessment can be viewed as a set of interlocking gears; they are all intricately connected to each other, and as you push on one, it sets the others in motion. Starting at any particular place of interest will eventually lead you into learning about the others.

Learning about formative assessment can take more than one productive path. Which path you take will depend on what you already have in place in your classroom and in your instructional repertoire, as well as where you *want* to go next. You can follow the chapters in this book for one path, but other than our contention that a focus on writing and using learning intentions and success criteria is an important starting place, we encourage you to feel free to alter that path.

Along the path of learning to implement the aspects of formative, keep in mind the following questions that sit at the heart of understanding the different aspects:

- *How do you convey important mathematics goals to students? How do you help them understand what it will look like when their learning is on track?*
- *How do you involve students more fully in evaluating and communicating about their own learning?*
- *How do you gather and interpret evidence of students' understanding?*
- *How do you use the results to determine appropriate responsive actions?*
- *How do you provide effective feedback to students to move their learning forward? And how do you help students learn to use feedback effectively?*

> **Recommendation 4:** Keep building your knowledge of how students learn and make sense of mathematics, particularly around their conceptual understanding of mathematics and the nature of persistent student difficulties and misconceptions.

All the various aspects of formative assessment rest not only on a thorough knowledge of the mathematics your students need to learn but also on an understanding of the typical barriers and difficulties that students have and the conceptual foundations of those difficulties. Because this knowledge is so fundamental to all of formative assessment, we found that it was a natural starting place for people's work in formative assessment.

Lesson Learned: Focusing on Students' Mathematical Understanding Is a Valuable Place to Start

We found that a good starting place for teachers' learning of formative assessment practices was to dig deeper into a question that teachers ask themselves regularly: *What is it that I want my students to understand?*

Our work focused initially on this question as an important starting place, because everything else a teacher does in formative assessment is based on how he or she articulates the important mathematics learning to himself or herself and to students. As we worked with teachers, one of the important distinctions that arose was between asking "What is it that I want students to *do*?" and asking "What is it that I want students to learn or understand *as a result of what they do*?" For example, in the early stages of our work together, teachers would often respond to the question, "What do I want my students to understand?" by saying, "I want my students to understand how to find the area of a triangle" or "understand how to graph a line"; the goals that they articulated to students were focused almost primarily on how-tos. Even those teachers who had a great deal of mathematics content knowledge found it easier to revert to focusing students on how-to, at least in the early writing of their learning intentions. Procedural goals were far easier to articulate than truly conceptual ones.

In response, we focused conversations on the distinction between describing the procedures that students master and describing the related important conceptual learning connected to those procedures. For example, if students needed to learn how to find the mean, the median, and the mode, we ended up talking about the importance of using statistics to make generalizations about a population. If students needed to learn how to find the area of a triangle, we ended up talking about the concept of area versus perimeter or volume, the reason for and use of formulas, and how rearranging a rectangle into a triangle could help you think about developing the area of a triangle from the area of a rectangle. Certainly the teachers deeply valued these goals of developing conceptual understanding as well as mastery of skills, and they were already looking for this level of understanding in their students, but we found that in many cases, it was *the clear articulation to students* of that conceptual learning that was lacking in their classrooms.

We felt that some work with unit progressions would support the teachers' articulation of that conceptual understanding. That work included identifying the key understandings (from a learning progression or other teaching resources) and breaking these into smaller concepts and subconcepts and then determining how those concepts connect and developed for students over the course of the unit. As teachers analyzed and constructed different unit progressions together, they developed a greater awareness of the various mathematical understandings that might underlie familiar math topics and were able to use this information to inform the development of learning intentions and success criteria.

TEACHERS' VOICES

In order to write quality learning intentions and success criteria, I need to get down and dirty with the content and find the underlying concepts and important math ideas. I also need to know where I am headed with instruction in order to help students see the targets also.

Because the use of learning intentions and success criteria is the foundation for all other formative assessment practices, as a starting point for your work in formative assessment, we encourage you to

- take the time to think about (or better yet, talk with colleagues about) an upcoming mathematics topic or unit of study and what you want students to understand mathematically as a result of learning various skills or procedure;

- find and read information about how students make sense of mathematical ideas to fuel those conversations; and
- use this information to work on developing unit progressions, learning intentions, and success criteria.

Starting this recommendation with the phrase *keep building your knowledge* is purposeful. Ongoing work to learn more about the development of students' mathematical understanding, particularly work with learning progressions, can support your development of unit progressions. Using and building learning progressions, as described in Chapter 6, includes identifying the key understandings, articulating how those understandings develop for students over the course of the unit, and determining some of the predictable misconceptions and stumbling blocks for students.

As you continue to write learning intentions and success criteria, we encourage you to continually refresh both your own knowledge base about how students learn and make sense of mathematics and to increase the repertoire of teaching strategies you have to meet the range of diverse learners. There are a number of resources in the field currently that can help you do this.

Reference Resource: Go to Resources.Corwin.com/
CreightonMathFormativeAssessment to access this resource.

Books and Articles List. This printable file provides a list of potential resources about how students learn.

Recommendation 5: Using formative assessment takes practice! Use it frequently so that it becomes habitual.

Throughout the earlier chapters, we introduced both planning guidelines and classroom resources, some of which are intended to provide you and your students with structures to practice and use important elements of formative assessment. For example, in Chapters 2 through 5, we referred to the "Framework for Using LI and SC (Learning Intentions and Success Criteria) with Students," which lays out a routine for sharing and revisiting learning intentions and success criteria across the span of a lesson. Accompanying that framework, there are various strategies suggested in Chapters 2 through 5 for sharing your learning intentions and success criteria with students and other strategies for revisiting them both in the middle of a lesson and at the conclusion of a lesson. (You may want to refer to the Appendix A for a summary of all the resources provided on the companion resource website.) These resources are important especially for your students, as they can help students learn to do *their* work of implementing formative assessment by giving them predictable steps to use. For formative assessment strategies and approaches to become habitual for you and your students, it's important to see them not as occasional activities but as a way to frame instruction and to incorporate them often enough to become familiar.

Initially, some of the teachers in the FACETS project were reluctant to start incorporating ideas in their classroom practices until they felt they clearly understood the critical aspect and had a clear picture how to use it with students. It's best to learn these formative assessment practices by trying them out with your students, and we encourage you to grant yourself permission to jump right in and try something new.

Lesson Learned: It's OK to Try Things Even When They're Not Perfect

The best way to learn to implement formative assessment practices is to learn by doing, yet it can sometimes feel daunting to try a new approach in a lesson when you have real pressures about getting through material with your students. In particular, some of the teachers were anxious about their initial attempts to write a learning intention and accompanying success criteria for a lesson, and they would spend time trying to ensure that the learning intention was written well, trying to find the perfect wording for the ideas in their head. Other teachers spent time thinking about ways to try to provide feedback within the course of the lesson.

We encourage efforts to carry out the practices well and according to the guidelines we've described; we also know that teachers have busy lives so we also simultaneously encourage realistic expectations for your planning time. A favorite mantra we had with some of the teachers in the FACETS project in relation to writing and sharing learning intentions, parodying the message at the end of some movies, was "No students will be harmed in the sharing of this learning intention," even when things aren't perfect—and in fact, many will probably be helped!

However, using formative assessment practices frequently does not necessarily mean you have to do them every day. You will need to decide for yourself what the right balance is for you between lessons that integrate formative assessment practices that are new to you, and lessons that don't. That balance will be different for each person. As these practices gradually become more familiar, it will become less of a question of whether or how to integrate them.

> **Recommendation 6: Include planning for formative assessment as part of your regular lesson planning.**

Most lesson planning involves thinking about a common set of questions:

- *What should students learn?*
- *What will they do to achieve that learning?*
- *How will that learning be evaluated?*
- *What do you do next as a result of what they have learned (or not learned)?*

Planning for formative assessment, however, involves adopting a slightly different mind-set that focuses on using the learning intentions and success criteria in a purposeful way and also on bringing students into the process of answering those questions. In the FACETS project, we provided teachers with a set of lesson-planning questions to support their efforts to create lessons with formative assessment practices embedded within them. We've provided two versions of a lesson-planning tool for you to consider; you may prefer one over the other.

Planning Resources: Go to Resources.Corwin.com/ CreightonMathFormativeAssessment to access these resources.

Guiding Questions for Formative Assessment. This is set of guiding questions to help you plan for incorporating formative assessment practices into a lesson.

Formative Assessment Planner Templates. This printable file provides two templates to help you create a formative assessment plan for a lesson.

Chapter 8 Summary Cards. These are index card-sized versions of the guiding questions, separated into planning and reflecting questions.

Recommendation 7: If some particular aspect of formative assessment is presenting an obstacle, turn your attention to a different one. Sometimes focusing on a different aspect can help untangle difficulties with another.

As we talked about previously, the various aspects of formative assessment are closely linked to each other. Much like interlocking gears, as you enact one aspect of formative assessment, you can see an effect on other aspects as well. Because of this close relationship among them, sometimes turning your attention to a different aspect and focusing your work there can help clarify confusing elements of a different aspect. The benefit of this relationship became clear to us in our work with teachers, leading to another important lesson learned: Deeper understanding about formative assessment (for you *and* your students) comes when you begin weaving the aspects together.

Lesson Learned: Deeper Understanding About Formative Assessment (for You and Your Students) Comes When You Begin Weaving the Aspects Together

Perhaps the most exciting development with the teachers in the FACETS project was when they began to move from viewing the various critical aspects as individual silos to understanding the interconnectedness of the aspects, and they started to get more mileage out of their implementation of formative assessment. For some, practices that had not been part of their instructional repertoire started becoming more seamlessly integrated. For others, instructional practices that they were already using took on a more specific purpose targeted to formative assessment. During one conversation after an observation, one teacher was excited to realize she had integrated *each* of the aspects into her lesson without really realizing it. (She had been focusing on others for her lesson that day.)

As teachers became more familiar with each of the aspects, they started being better able to turn their attention to weaving them together, and this second round of learning, which often took place in the empowering students phase (see Figure 8.2), resulted in a new series of "ahas!" for teachers. Recall the teacher at the start of the chapter who said,

> [F]or me, it took courage to keep going with these little itty bitty pieces, knowing that the little itty bitty pieces were making a huge difference even if I didn't have the global picture. And the other pieces snapped into focus for me over the time period I worked with it, very dramatically at very unexpected moments. And every time something has snapped into focus, the whole picture has gotten easier.

Whatever your entry point and your path through developing and honing your use of these formative assessment practices, we've found that *it's important not to dwell too long trying to master any single critical aspect*, because it is often in the work of trying to weave them together that deeper understanding of all of them results. There is certainly value in giving yourself a realistic learning curve and trying out any elements of this that are new to you in a very gradual way. During these times, it can be important to just focus on one piece at a time. However, after some focused attention to one aspect, it can also be fruitful to move on to explore a different aspect before you feel like you've fully mastered the current one. As you try to integrate one practice after another into a coherent approach to instruction, new insights invariably will arise.

TEACHERS' VOICES

I was kind of overwhelmed at first with writing learning intentions and success criteria, especially when we started moving forward into all of the other pieces of formative assessment. It was like, wait a second I haven't really mastered writing the learning intentions and success criteria yet, and we are moving on. But, I think the reason we were finally able to write learning intentions and success criteria so well this year is because we saw what the end result was going to be and how we were going to use the learning intentions and success criteria. By learning more about the other pieces of formative assessment and knowing how we were going to use the learning intentions and success criteria, we were able to write them better. So, it was okay that we were a little overwhelmed at first—we had to see everything else to get the big picture.

CONCLUSION: FINAL WORDS OF ENCOURAGEMENT ■

> Recommendation 8: Be kind to yourself; don't try to do it all at once.

Have we mentioned the importance of allowing yourself time to learn and practice? We feel this recommendation is pretty key, so we've made it our first and last recommendation for this chapter.

At the end of the FACETS professional development work, we asked some of the teachers what advice they would give someone who was starting out on their path toward full implementation of these practices. One sixth-grade teacher wrote,

> *Learning about formative assessment and implementing formative assessment are very different. It feels so familiar when you are learning about it, but then when you start to implement formative assessment, it starts to feel very messy or unfamiliar. Be patient with yourself, and do not be afraid to just try bits and pieces of the whole picture. Those pieces will start to connect and make sense as a whole. At times, I would feel like I understood formative assessment then something would happen to make me feel like I had no clue. I can still have both of those moments. Learning is a messy process, but when students have a clear learning intention and have a set of success criteria to go by and they take ownership of their work, it all starts working! Formative assessment makes you look at teaching and learning through a different lens: one where students are involved in their own learning.*

We'd like to underscore several things she says:

- Learning is a messy process, so sometimes the implementation of these practices will feel clear and other times, it may feel like you're back to not understanding how to help it happen in your classroom. Allowing yourself time and permission to take some steps forward and some steps backward can be a helpful mind-set as you work toward implementation.
- Do not be afraid to just try bits and pieces of the whole picture. In our experience with teachers, many of them needed to focus on a particular critical aspect, work with it to become more comfortable with it, and temporarily ignore the other critical aspects while they were developing this comfort.
- Hang in there! It does make sense, and it gets easier with time.
- Your own growth may surprise you. It certainly did for many of the teachers we worked with.

Your path to implementing formative assessment will be unique, as you make these practices your own. Your students are the ones who stand to benefit most from your efforts, and we hope that you find as much pleasant surprise, success, and new doors opened as many of the teachers in the FACETS project did. Bon voyage!

Recommendations in This Chapter

1. Be kind to yourself; don't try to do everything all at once.

2. There is benefit in collaboration! Seek support for this work.

3. Use a self-reflection tool to check in on your progress, set new goals, and adjust your path.

4. Keep building your knowledge of how students learn and make sense of mathematics, particularly around their conceptual understanding of mathematics and the nature of persistent student difficulties and misconceptions.

5. Using formative assessment takes practice! Use it frequently so that it becomes habitual.

6. Include planning for formative assessment as part of your regular lesson planning.

7. If some particular aspect of formative assessment is presenting an obstacle, turn your attention to a different one. Sometimes focusing on a different aspect can help untangle difficulties with another.

8. And again, be kind to yourself; don't try to do it all at once.

Reference Resources: Go to Resources.Corwin.com/ CreightonMathFormativeAssessment to access these resources.

 Chapter 8 Summary Cards. This printable file provdes a Summary Card of the recommendations in this chapter.

 Formative Assessment Recommendations. This printable file provdes a summary of the all recommendations in this book.

■ RESOURCES

The following sections present some resources that can help you implement the recommendations. Each of these resources is referenced in earlier sections of the chapter, but here we provide a consolidated list. All resources can be found at Resources.Corwin.com/CreightonMath FormativeAssessment.

Learning Resources

This resource supports reflecting about your implementation of formative assessment practices in order to consider next steps and focus:

- **Reflecting on Your Implementation** provides indicators by which you can assess the extent to which you are incorporating the principles into your instructional practices and consider next steps as you work gradually toward full implementation.

In addition, the videos on the website can help you reflect on your implementation efforts.

Reference Resources

These resources summarize key ideas about implementing formative assessment:

- **Implementation Principles for Formative Assessment** provides a list of the principles described in this chapter.
- **Implementation Indicators for Formative Assessment** includes the full list of the principles and associated indicators (also in Appendix B).
- **Correlation of the Principles, Recommendations, and Resources** provides the principles correlated with associated recommendations and resources.
- **Formative Assessment Cycle** is a color version of the *full* Formative Assessment Cycle.
- **Formative Assessment Cycle Video** is an audiovisual tour of the Formative Assessment Cycle.
- **Books and Articles List** is a bibliography of recommended sources for additional information, for yourself or others.
- **Chapter 8 Summary Cards** provide index-sized Summary Cards for the following:
  - **Planning for Formative Assessment** is a quick reference of the planning questions from the Guiding Questions for Formative Assessment.
  - **Reflecting on the Lesson** is a quick reference of the reflection questions from the Guiding Questions for Formative Assessment.
  - **Recommendations for Moving Toward Implementation** includes all the recommendations from this chapter.
- **Formative Assessment Recommendations** is a printable PDF that includes all the recommendations from the book.

Planning Resources

These planning tools are resources to support your lesson planning:

- **Guiding Questions for Formative Assesment** can help you include formative assessment practices in your lesson plan.
- **Formative Assessment Planner Templates** can be used for creating a formative assessment plan for a lesson.

Appendix A

Resources

The following sections present some resources that can help you implement the recommendations found in this book. Each of these resources is referenced in relevant chapters, but here we provide a consolidated list. All resources, plus additional videos, can be found at Resources.Corwin.com/ CreightonMathFormativeAssessment.

LEARNING RESOURCES ■

These resources support your learning about the critical and supporting aspects of formative assessment.

Guidance Documents

- **Reflecting on Your Implementation** provides indicators by which you can assess the extent to which you are incorporating the principles into your instructional practices and consider next steps as you work gradually toward full implementation.

Web-Based Interactives

- **Formative Assessment Overview** provides an approach to considering different characteristics of formative assessment. The interactive web page provides several statements about formative assessment, by selected notable authors, which you can sort to look for connections among the ideas.
- **Lesson Purposes, Practice Interactive** provides practice considering purposes for lessons when writing learning intentions and success criteria; it also helps you learn how to use the "Lesson Purposes, Planning Interactive" listed in the planning resources.
- **Responsive Actions—During** provides practice deciding which responsive action is reasonable, for individual students, small groups, or a whole class.
- **Responsive Actions—After** provides practice deciding which responsive action is reasonable after the lesson is concluded.

- **Feedback Characteristics—Content** provides practice deciding whether feedback examples meet the characteristics related to content.
- **Feedback Characteristics—Usability** provides practice deciding whether feedback examples meet the characteristics related to student usability.

■ IMAGES FROM PRACTICE

- **Examples of Learning Intentions and Success Criteria.** Although writing learning intentions and success criteria is dependent on many factors including curricula, unit goals, and data from previous lesson, many teachers have found reviewing examples helpful. This searchable database includes the examples used throughout this book as well as additional examples.
- **Images of Posted Learning Intentions and Success Criteria** shows various ways teachers have posted learning intentions and success criteria for reference and use during a lesson.
- **Sample Unit Progressions** we created collaboratively with teachers in our professional development program.
- **Modeling Reasoning and Approach Example Video** portrays a teacher modeling how to share your reasoning and approach to a mathematics problem.

■ EXTERNAL RESOURCES

- **Books and Articles List** is a bibliography of recommended sources for additional information, for yourself or others.

■ REFERENCE RESOURCES

These resources summarize key ideas about formative assessment practices presented in this book.

■ SUMMARY CARDS

- **Chapter 1 Summary Card** is an index-sized Summary Card that provides a summary of the critical and supporting aspects of formative assessment.
- **Chapter 2 Summary Cards** provides index-sized Summary Cards for the following:

  o **Characteristics of Learning Intentions and Success Criteria** summarizes the characteristics described in this chapter.
  o **Learning Intentions and Success Criteria Starter Statements** includes a list of sentence starters based on using *understand* as the basis of the learning intention.
  o **Lesson Purposes** lists the lesson purposes described more fully in the "Overarching Purposes and Guidelines for Writing LI and SC."
  o **Framework for Using LI and SC With Students** summarizes ways to help students understand and use learning intentions and

success criteria during a lesson. The framework is described in detail in the classroom resource listed next.

  o **Recommendations for Using Mathematics Learning Intentions and Success Criteria** includes all the recommendations from Chapter 2.

- **Chapter 3 Summary Cards** provides index-sized Summary Cards for the following:

  o **Planning for Eliciting Evidence** summarizes considerations when planning for eliciting evidence.
  o **Process for Eliciting and Interpreting Evidence** summarizes evidence related steps from planning for the lesson to enacting the lesson to determining next steps after the lesson.
  o **Responsive Actions** summarizes four responsive actions.
  o **Recommendations for Gathering, Interpreting, and Acting on Evidence** includes all the recommendations from Chapter 3.

- **Chapter 4 Summary Cards** provides index-sized Summary Cards for the following:

  o **Characteristics of Formative Feedback** summarizes the characteristics described in this chapter.
  o **Recommendations for Providing and Using Formative Feedback** includes all the the recommendations from Chapter 4.
- **Chapter 5 Summary Card** is an index-sized Summary Card that includes all the recommendations described in Chapter 5.
- **Chapter 6 Summary Cards** provides index-sized Summary Cards for the following:

  o **Building a Unit Progression** summarizes the steps involved in building a unit progression.
  o **Using a Unit Progression to Write LI and SC** summarizes the steps described in the guideline document in Planning Resources.
  o **Recommendations for Using Mathematics Learning Progressions** includes all the recommendations from Chapter 6.

- **Chapter 7 Summary Cards** provides index-sized Summary Cards for the following:

  o **Considerations for Environment** includes brief descriptions of each of the elements described in Chapter 7.
  o **Recommendations for Establishing a Classroom Environment** includes all the recommendations from Chapter 7.

- **Chapter 8 Summary Card** is an index-sized Summary Card that includes all the recommendations from Chapter 8.

GUIDING DOCUMENTS ■

- **Formative Assessment Recommendations** is a printable PDF that includes all the recommendations from this book.
- **Implementation Principles for Formative Assessment** provides a full list of the principles.
- **Implementation Indicators for Formative Assessment** includes the full list of the principles and associated indicators (also in Appendix B).

- **Correlation of the Principles, Recommendations, and Resources** provides the principles correlated with associated recommendations and resources.

■ FORMATIVE ASSESSMENT CYCLE

- **Formative Assessment Cycle** is a color version of the *full* Formative Assessment Cycle that we build throughout this book.
- **Formative Assessment Cycle Video** is an audiovisual tour of the Formative Assessment Cycle that we build throughout this book.

■ PLANNING RESOURCES

These planning tools are resources to support your lesson planning when writing learning intentions and success criteria for your mathematics lessons.

■ UNIT-LEVEL PLANNING

- **Building a Unit Progression** includes a step-by-step process for building a unit progression.
- **Unit Progression Builder** is an interactive web page that uses the process from the "Building a Unit Progression" resource to help you build a progression for any unit you are planning.

■ LESSON-LEVEL PLANNING

- **Guiding Questions for Formative Assessment** includes guiding questions to help you plan for formative assessment practices.
- **Formative Assessment Planner Templates** can be used for creating a formative assessment plan for a lesson.
- **Guidelines for Writing LI and SC** includes a step-by-step process for writing learning intentions and success criteria after you have created a unit level progression.
- **Evaluating and Refining LI and SC** provides questions designed to help you evaluate and refine your learning intentions and success criteria. These questions are also included in the "Guidelines for Writing LI and SC" resource.
- **Lesson Purposes, Planning Interactive** provides a structure (similar to the "Lesson Purposes, Practice Interactive") for selecting purposes for lessons within a unit.
- **Overarching Purposes and Guidelines for Writing LI and SC** is a printable guide for considering purposes for lessons when writing learning intentions and success criteria. It can also be used as a reference when using the "Lesson Purposes, Planning Interactive."
- **Using Exit Tickets as Evidence** describes a way to sort exit tickets to inform your work for the next day.

CLASSROOM RESOURCES ■

These resources illustrate various classroom routines that you can use during instruction. Each routine provides a structure that you can use or adapt to routinize your practice around the use of learning intentions and success criteria.

STRATEGIES AND TECHNIQUES ■

- **Framework for Using LI and SC With Students** describes a way to help students understand and use learning intentions (LI) and success criteria (SC) during a lesson. It describes actions for both teacher and students to take at the start of the lesson, at midway points during the lesson, and at the end of the lesson.
- **Sharing Strategies** includes several strategies for use when introducing students to LI and SC for the first time.
- **Revisiting Strategies** includes several strategies for use when revisiting LI and SC throughout the lesson.
- **Wrapping-Up Strategies** includes several strategies for use referring to LI and SC to wrap up a lesson.
- **Teacher Summary Cards for the Strategies** is a set of large-sized cards that provides step-by-step summaries of *teacher* moves for the strategies described earlier.
- **Elicitation Strategies** includes a description of elicitation techniques.
- **Feedback Techniques** lists formative assessment classroom techniques (FACTS) related to feedback that can be found in the book *Mathematics Formative Assessment: 75 Practical Strategies for Linking Assessment, Instruction, and Learning* by Keeley and Tobey (2011).
- **Feedback Response Strategy** describes a strategy that you can give your students to help them respond to feedback.
- **Self-Assessment Techniques** lists FACTS for student self-assessment that can be found in the book *Mathematics Formative Assessment: 75 Practical Strategies for Linking Assessment, Instruction, and Learning* by Keeley and Tobey (2011).

LESSONS TO TEACH FORMATIVE ASSESSMENT PRACTICES ■

- **Introducing Students to Learning Intentions and Success Criteria** is an introductory lesson to help students understand the purpose of learning intentions and success criteria.
- **Introduction to Formative Feedback (for Students)** is a lesson plan for introducing the characteristics of formative feedback to students.
- **Peer Feedback** is a lesson plan for introducing and practicing how to give formative feedback to peers.

CLASSROOM MATERIALS ■

These resources can be copied or re-created to be used in your classroom.

■ STUDENT SUMMARY CARDS

- **Student Summary Cards for the Strategies** is a set of large-sized cards that provides step-by-step summaries of *student* moves for the strategies described with the Classroom Resources. These cards can be distributed to each student or pairs of students for use when learning about a strategy.
- **Student Summary Card for Participating as a Listener** is a large-sized card that provides a summary for *students* for the listening strategy in Chapter 7.

■ TEMPLATES

- **Feedback Templates** for using the feedback techniques described above.

■ POSTERS

- **Self-Regulation Poster** can be posted in your classroom to encourage students to think like a self-regulating learner.
- **Self-Assessment Poster** can be posted in your classroom to provide prompts that encourage students to provide evidence from their work in their self-assessment.
- **Feedback Poster** can be posted in your classroom for ongoing reference after introducing students to the characteristics of feedback through the "Introduction to Formative Feedback (for Students)" lesson. The file can be printed poster size though your local office supply or copy center.

Appendix B

Implementation Indicators for Formative Assessment

LEARNING PROGRESSIONS ■

Implementation Principle A: Be clear about what's mathematically important or significant in your content (the concepts as well as the procedures) and how that content connects with other mathematics and grows both within a unit and across units.

- I use learning progressions to help inform my thinking about the mathematics content I will teach.

LEARNING INTENTIONS AND SUCCESS CRITERIA ■

Implementation Principle B: Learning intentions help your students focus on what they should be learning, and success criteria help them gauge the extent to which they are learning it.

- My learning intentions make clear what the central focus of learning is for the lesson.

- My learning intentions are about conceptual understanding.

- For each learning intention, I include success criteria that focus on students demonstrating their understanding through explanations, justifications, or other higher-order thinking activities, as well as those that focus on completing procedures correctly.

- My success criteria focus on evidence that students can tangibly demonstrate.

- My success criteria for a learning intention, taken collectively, provide enough evidence for my students and me to be confident whether the learning intention has been reached.

- I write learning intentions and success criteria whose language and math content are appropriately accessible to students.

- The lesson activities provide experiences that enable students to make progress toward the learning intention.

Implementation Principle C: You and your students both need access to the learning intention and success criteria throughout the lesson.

- I post the learning intentions and success criteria, or provide them on a handout, so that my students and I can refer to them throughout the lesson.

■ ELICITING AND INTERPRETING EVIDENCE

Implementation Principle D: Students need opportunities and encouragement to articulate the reasoning behind their work.

- I provide opportunities for students to articulate the reasoning behind their work, as it relates to the success criteria, for correct as well as incorrect work.

Implementation Principle E: Students need opportunities to provide evidence of each of the success criteria.

- When I plan my lessons, I think about the kind of evidence I need, with regard to my learning intention and success criteria.

- My lessons include opportunities for students to provide evidence for each of the success criteria.

Implementation Principle F: You need variety in how, when, and from whom you gather evidence, to get a full picture of the class's understanding, as a whole.

- I collect evidence from either a few individuals, small groups, or the whole class, as appropriate at given points in a lesson

- I collect evidence at key points throughout the lesson (not just at the end).

- I collect evidence through student explanations, justifications, and so on, as well as through their completion of procedures and problems.

Implementation Principle G: All gathering and interpretation of evidence needs to be in terms of the success criteria.

- I focus on the success criteria when I assess the current status of students' progress.

- I choose and carry out a responsive action *during instruction*, based on evidence I have gathered during the lesson (e.g., use evidence to group students for a follow-up activity or to back up to provide more instruction).

- I use evidence gathered from a lesson to plan a responsive action *for the next day's lesson*.

FORMATIVE FEEDBACK ■

Implementation Principle H: Formative feedback communicates to the student how he or she has met, as well as not yet met, the success criteria, and it provides hints, cues, or models for next steps.

- I provide feedback that includes what has been met (including what in the work indicates this to me).

- I provide feedback that includes what hasn't been met (and why not, when appropriate).

- I provide feedback that includes hints, cues, or models—not solutions or answers—to help students meet the success criteria.

Implementation Principle I: Formative feedback should be focused and specific.

- I focus my feedback on the skills and concepts described in the success criteria.

- My feedback to students focuses on no more than one or two things at a time, so the students can digest it and act on it.

Implementation Principle J: You need to prepare yourself to provide formative feedback and to plan opportunities for students to respond to it.

- During my lesson planning, in order to help me plan for providing feedback, I articulate for myself what a student would say or do if he or she is meeting the success criteria.

- I provide structures and time for students to act on feedback.

- I follow-up with students to see if they acted on feedback (during the lesson or the next day).

STUDENT OWNERSHIP AND INVOLVEMENT ■

Implementation Principle K: Students need to understand what the learning intention and success criteria mean.

- I discuss learning intentions and success criteria and make efforts to clarify their meanings.

- I share and discuss sample work against success criteria, as a way to build understanding of the success criteria and what it looks like to meet them.

- I refer to learning intention or success criteria (verbally or by pointing to a displayed version) to refocus students or help them solidify their understanding.

- I return to the learning intention and success criteria as part of lesson closure.

Implementation Principle L: Students need to use the success criteria to guide their self-assessment and self-monitoring.

- I provide opportunities for my students to self-assess according to the success criteria.

- I provide structures and strategies designed to help students monitor their own learning.

- My students refer to the success criteria to help focus and guide their work.

- When giving feedback to a peer, my students refer to the learning intention or success criteria.

Implementation Principle M: Students need to learn how to provide and use evidence of their mathematical reasoning and understanding in their self-assessment, and they need encouragement to do so.

- My students supply evidence of their work, with a comparison to the success criteria, in their self-assessments.

- My students ask questions that are focused on moving their learning forward (i.e., are more specific than "I don't get it" and focus on a particular element of what they are learning).

Implementation Principle N: Students need to understand what feedback is, why it's helpful, and how they should use it.

- I spend time in class teaching students what feedback is and how they will use it.

- I build in opportunities to check that students understand the feedback they are receiving (for example, summarizing it in their own words or telling me what they think they need to do next).

- My students use teacher and peer feedback to improve their work. They may do so with prompting from me but also sometimes without prompting from me.

Implementation Principle O: Students need opportunities to provide and receive peer feedback, as a way to learn to provide feedback to themselves about their own work.

- I provide opportunities for students to practice giving peer feedback.

- I help students connect the process of peer feedback with the process of self-assessment.

Implementation Principle P: Students understand a variety of ways to act on the results of their self-monitoring to take the next steps to move their learning forward.

- My students access resources (including peers, print resources, and the teacher) to help them make progress toward meeting the success criteria. They may do so with prompting from me but also sometimes without prompting from me.

CLASSROOM ENVIRONMENT ■

Implementation Principle Q: To support formative assessment practices, the social/cultural environment of the class promotes intellectual safety and curiosity.

- I establish clear guidelines for my students about everyone communicating their ideas safely, and I use structures during the lesson to ensure this happens.

- I provide appropriate think time after a question or comment before continuing.

- I ask questions about students' thinking, both for correct thinking as well as incorrect thinking.

Implementation Principle R: To support formative assessment practices, the instructional environment optimizes learning and encourages and makes visible students' thinking.

- When I plan lessons, I think about framing my math activities in a way that invites my students to do some mathematical thinking and discussion.

- I try to provide just enough support and information so that students can do the majority of the work themselves.

Implementation Principle S: To support formative assessment practices, the physical environment allows easy access to various resources.

- I have a predictable location for making available my learning intentions and success criteria, so students always know where to look.

- I arrange my classroom to give students ready access to the formative assessment tools they need as well as each other.

References

Andrade, H. L. (2010). *Students as the definitive source of formative assessment: Academic self-assessment and the self-regulation of learning.* Paper presented at NERA Conference Proceedings Rocky Hill, CT. Retrieved from http://digital commons.uconn.edu/nera_2010/25

Behr, M., Harel, G., Post, T., & Lesh, R. (1992). Rational number, ratio, and proportion. In D. Grouws (Ed.), *Handbook of research on mathematics teaching and learning* (pp. 296–333). New York, NY: Macmillan.

Black, P., & Wiliam, D. (1998). Assessment and classroom learning. *Assessment in Education: Principles, Policy & Practice, 5*(1), 7–73.

Bright, G., & Joyner, J. (2004). *Dynamic classroom assessment: Linking mathematical understanding to instruction.* Vernon Hills, IL: ETA/Cuisenaire

Brookhart, S., Moss, C., & Long, B. (2008). Formative assessment that empowers. *Educational Leadership, 66*(3), 52–57.

Center on Instruction. (2008). *A synopsis of Hattie & Timperley's "Power of Feedback."* Portsmouth, NH: RMC Research Corporation: Author.

Common Core Standards Writing Team. (2013). *Progressions for the Common Core State Standards in Mathematics (draft): Front matter, preface, introduction.* Tucson, AZ: Institute for Mathematics and Education, University of Arizona. Retrieved from http://ime.math.arizona.edu/progressions/

Council of Chief State School Officers. (2008). *Attributes of effective formative assessment.* Washington, DC: Author. Retrieved July 15, 2013 from http://www.ccsso.org/Resources/Publications/Attributes_of_Effective_Formative_Assessment.html.

Cramer, K., & Post, T. (1993). Connecting research to teaching proportional reasoning. *Mathematics Teacher, 86*(5), 404–407.

Danielson, C. (2007). *Enhancing professional practice: A framework for teaching, 2nd edition.* Alexandria, VA: ASCD.

Driscoll, M. (1999). *Fostering algebraic thinking: A guide for teachers grades 6–10.* Portsmouth, NH: Heinemann.

Hattie, J., & Timperley, H. (2007). The power of feedback. *Review of Education Research, 77,* 81–112.

Heritage, M. (2010). *Formative assessment: Making it happen in the classroom.* Thousand Oaks, CA: Corwin.

Heritage, M. (2011). *Developing learning progressions.* Paper presented at the Annual Meeting of the American Educational Research Association, New Orleans, LA.

Heritage, M., Kim, J., Vendlinski, T., & Herman, J. (2009). From evidence to action: A seamless process in formative assessment? *Educational Measurement: Issues and Practice, 28*(3), 24–31.

Institute for Mathematics and Education. (2007). *Progression documents for the common core math standards.* Retrieved from http://ime.math.arizona.edu/progressions/

Jacob, B., & Fosnot, C. T. (2008). *Best buys, ratios, and rates: Addition and subtraction of fractions.* Portsmouth, NH: Heinemann.

Keeley, P., & Tobey, C. (2006). *Mathematics curriculum topic study: Bridging the gap between standards and practice.* Thousand Oaks, CA: Corwin.

Keeley, P., & Tobey, C. (2011). *Mathematics formative assessment: 75 practical strategies for linking assessment, instruction, and learning.* Thousand Oaks, CA: Corwin.

Kilpatrick, J., Swafford, J., & Findell, B. (Eds.). (2001). *Adding it up: Helping children learn mathematics.* Washington DC: National Academy Press.

Lamon, S. (1994). Ratio and proportion: Cognitive foundations in unitizing and norming. In J. Confrey (Ed.), *The development of multiplicative reasoning in the learning of mathematics* (pp. 89–122). Albany: State University of New York Press.

Leahy, S., Lyon, C., Thompson, M., & Wiliam, D. (2005). Classroom assessment: Minute by minute, day by day. *Educational Leadership, 63*(3), 19–24.

Lee, C. (2006). *Language for learning mathematics: Assessment for learning in practice.* New York, NY: Open University Press.

Lesh, R., Post, T., & Behr, M. (1988). Proportional reasoning. In J. Hiebert & M. Behr (Eds.), *Number concepts and operations in the middle grades* (pp. 93–118). Reston, VA: Lawrence Erlbaum & National Council of Teachers of Mathematics.

Masters, G., & Forster, M. (1997). *Developmental assessment.* Victoria: The Australian Council for Educational Research.

National Governors Association Center for Best Practices & Council of Chief State School Officers. (2010). *Common Core State Standards Mathematics.* Washington, DC: Author.

Nicol, D. J., & Macfarlane-Dick, D. (2006). Formative assessment and self-regulated learning: A model and seven principles of good feedback practice. *Studies in Higher Education, 31,* 199–218.

Popham, W. J. (2008). *Transformative assessment.* Alexandria, VA: ASCD.

Rational Number Project. (2002). *Chronological bibliography.* Retrieved from http://www.cehd.umn.edu/ci/rationalnumberproject/bib_chrono.html

Reinhart, S. C. (2000). Never say anything a kid can say! *Mathematics Teaching in the Middle School, 5*(8), 478–483.

Rowe, M. (1986). Wait time: Slowing down may be a way of speeding up! *Journal of Teacher Education, 37*(1), 43–50.

Sadler, R. D. (1989). Formative assessment and the design of instructional systems. *Instructional Science, 18,* 119–144.

Stiggins, R. (2007). Assessment through student eyes. *Educational Leadership, 64*(8), 22–26.

Student Achievement Division, Ontario Ministry of Education. (2011). *Capacity building series: Asking effective questions in mathematics.* Retrieved from http://www.edu.gov.on.ca/eng/literacynumeracy/inspire/research/capacitybuilding.html

Wiggins, G. (2012). Seven keys to effective feedback. *Educational Leadership, 70*(1), 10–16.

Wiliam, D. (1999). Formative assessment in mathematics—part 2: Feedback. *Equals: Mathematics and Special Educational Needs, 5*(3), 8–11.

Wiliam, D. (2000). Formative assessment in mathematics part 3: The learner's role. *Equals: Mathematics and Special Educational Needs, 6*(1), 19–22.

Wiliam, D. (2011). *Embedded formative assessment.* Bloomington, IN: Solution Tree Press.

Wilson, M. R., & Bertenthal, M. W. (Eds.). (2005). *Systems for state science assessment.* Board on Testing and Assessment, Center for Education, National Research Council of the National Academies. Washington, DC: National Academies Press.

Index

Actions, responsive, 90–99
  formative feedback as, 128–130
  helping students learn about, 181–183
  by students receiving formative
    feedback, 147–148
Adding It Up, 29
Andrade, H., 6, 12

Brookhart, S., 12

Card sorts, 75
Clarification Strategy, 169
Classroom environment, 11, 217–218, 281
  elements of, 218–222
  evidence-gathering materials in, 241
  implementation principles for formative
    assessment and, 253
  instructional, 218–219, 234–237
  intellectual curiosity and, 229–232
  intellectual safety and risk-taking in,
    223–229
  keeping resources available in, 237–240
  learning materials in, 240
  math tasks in, 234–236
  peers as resources in, 242–243
  physical, 219, 237–240
  self-assessment materials in, 241–242
  social/cultural, 218, 222–223, 233
  students' responsibility for doing the
    learning in, 236–237
Classroom(s)
  materials, 62, 108, 151, 190, 245, 275–276
  resources, 61–62, 107–108, 150–151,
    189–190, 274–276
Collaboration, 231–232
  by teachers, 258–260
Common Core State Standards, 29–30
  learning progressions and, 199
  sources of evidence and, 69
Council of Chief State School Officers, 5
Countdown timers, 229
Cover-Up Strategy, 169
Criteria, success, 7–8
  additional key characteristics of, 31–36
  aligned to learning intention, 34–35
  aligning evidence to, 69–71

characteristics of, 27–36
clarification of goals using, 158–160
common pitfalls to avoid with, 50–52
creating, 36–58
defined, 24–27
demonstrating reaching the learning,
    46–52
evidence-gathering saved for end of
    class and, 76–80
formative feedback providing guidance
    on what student needs to do next to
    meet, 118–119, 136–141
formative feedback referencing extent of
    meeting, 117
helping students learn about, 167–175
implementation indicators, 277–281
implementation principles for formative
    assessment and, 249
as indication of reaching learning
    intention, 33–34
interpreting evidence of meeting, 86–90
lesson planning with focus on, 38–52
not always "leveled," 51
resources, 275
self-assessment and student ownership,
    160–162
sounding like the lesson agenda, 50
tangible or observable examples, 32–33
teacher role in sharing and using, 53–58
teachers supporting students' use of,
    56–58
as too specific, 50
turning into a long list, 50–51
wording of, 44–46
written to be understandable by
    students, 35–36
  See also Learning intentions
Critical aspects of formative assessment,
    7–11
Curiosity, intellectual, 229–232
  intellectual safety and, 222–223
Cycle, formative assessment, 12–15

Danielson, C., 217
Demonstrating reaching the learning,
    46–52

Dependent and independent variables, 127
Diagnosing student difficulties using
  learning progressions, 210–213

Elicitation of evidence
  implementation principles for formative
    assessment and, 250
  during lessons, 81–86
  misaligning task with, 73–76
  planning for eliciting, 72–80
  students' role in, 102–105
  See also Evidence
Elicitation tasks, 65
Eliciting and interpreting evidence.
    See Evidence
Environment, classroom,
    11, 217–218, 281
  elements of, 218–222
  evidence-gathering materials in, 241
  implementation principles for formative
    assessment and, 253
  instructional, 218–219, 234–237
  intellectual curiosity and, 229–232
  intellectual safety and risk-taking in,
    223–229
  keeping resources available in, 237–240
  learning materials in, 240
  math tasks in, 234–236
  peers as resources in, 242–243
  physical, 219, 237–240
  self-assessment materials in, 241–242
  social/cultural, 218, 222–223, 233
  students' responsibility for doing the
    learning in, 236–237
Evidence, 8–9
  aligned to learning intentions and
    success criteria, 69–71
  choosing a responsive action to, 90–99
  defined, 64–68
  elicitation during lessons, 81–86
  in formative assessment cycle, 12–15
  -gathering materials, 241
  gathering saved for end of class, 76–80
  helping students learn about, 175–178
  implementation indicators, 278–279
  implementation principles for formative
    assessment and, 250
  interpretation for individual
    students, 80
  interpretation for meeting success
    criteria, 86–90
  managing information from, 100–101
  misaligning task with, 73–76
  planning for eliciting, 72–80
  self-assessment and student ownership,
    160–162
  sources of, 68–69
  students' role in elicitation of, 102–105
  using questions to jump to instruction
    and, 84–86

Exit tickets, 107, 173
Expressions for figure patterns, 69
External resources, 272

Feedback, formative, 6, 9–10
  after lessons, 96–98
  characteristics of, 115–123
  concrete, 120
  dangers of hit-and-run feedback and,
    142–143
  defined, 110–115
  drafting possible model response for,
    130–132
  example, 109–110
  formative, 9–10
  helping students learn about, 178–181
  implementation indicators, 279
  implementation principles for formative
    assessment and, 251
  during lessons, 91
  planning for, 130–134
  planning how students might act on, 134
  planning the way to give, 133
  posters, 190
  prioritized, 120–121
  providing guidance on what student
    needs to do next to meet success
    criteria, 118–119
  as responsive action, 128–130
  self-assessment and student ownership,
    160–162
  student responses to, 147–148
  student's role in, 143–148
  students understanding, 144–146
  teacher's role during lesson and,
    134–143
  templates, 190
  timely, 121–122
  as usable by the student, 120–122
  whole-group versus individual, 123–128
Formative assessment
  compared to teaching practices, 16–17
  critical aspects of, 7–11
  definition of, 5–7
  developing self-regulation skills
    through, 5–11
  focus on students' mathematical
    understanding as start in, 261–264
  giving yourself permission to focus
    on your own learning first and,
    255–257
  implementation principles for, 249–253
  learning by doing in, 264–265
  putting students front and center, 15–16
  supporting aspects of, 11
  sustained over the long term, 254–266
  taking different paths in learning to
    implement, 260–261
  taking time for, 257–258
  teachers collaborating on, 258–260

teaching students how to participate in, 18–20
used in the classroom, 12–17
weaving aspects together for deeper understanding about, 265–266
Formative assessment cycle, 12–15, 25, 26 (figure), 67 (figure), 247–249
See also Evidence
Formative Assessment in the Mathematics Classroom: Engaging Teachers and Students (FACETS), 254, 267
Formative feedback, 6, 9–10
after lessons, 96–98
characteristics of, 115–123
concrete, 120
dangers of hit-and-run feedback and, 142–143
defined, 110–115
drafting possible model response for, 130–132
example, 109–110
helping students learn about, 178–181
implementation indicators, 279
implementation principles for formative assessment and, 251
during lessons, 91
planning for, 130–134
planning how students might act on, 134
planning the way to give, 133
posters, 190
prioritized, 120–121
providing guidance on what student needs to do next to meet success criteria, 118–119
as responsive action, 128–130
self-assessment and student ownership, 160–162
student responses to, 147–148
student's role in, 143–148
students understanding, 144–146
teacher's role during lesson and, 134–143
templates, 190
timely, 121–122
as usable by the student, 120–122
whole-group versus individual, 123–128

Gathering Evidence to Learn More, 172
Goals
clarification of, 158–160
taking action toward, 165–166
Guidance documents, 271, 273

Heritage, M., 12, 217
Highest-priority learning, 31–32, 40–46
Hinge points, 75–76
Hit-and-run feedback, 142–143

Images from practice, 272
Implementation principles for formative assessment, 249–253
Independent and dependent variables, 127
Individualized Education Programs (IEPs), 213
Information management, evidence, 100–101
Instructional environment, 218–219
framed to encourage and make visible students' thinking and to optimize learning, 234
math tasks and, 234–236
students' responsibility for doing the learning in, 236–237
Intellectual curiosity, 229–232
intellectual safety and, 222–223
Intellectual safety
curiosity and, 222–223
risk-taking and, 223–229
Intentions, learning, 7–8
additional key characteristics of, 31–36
aligned to success criteria, 34–35
aligning evidence to, 69–71
characteristics of, 27–36
clarification of goals using, 158–160
creating, 36–58
defined, 24–27
and demonstrating reaching the learning, 46–52
focused on learning, not activities, 27–31
focusing lesson on highest-priority learning for that lesson, 31–32, 40–46
helping students learn about, 167–175
implementation indicators, 277–281
implementation principles for formative assessment and, 249
lesson planning with focus on, 38–52
resources, 275
revisiting, 174–175
teacher role in sharing and using, 53–58
teachers supporting students' use of, 56–58
teacher use of, 53–55
wording of, 44–46
written to be understandable by students, 35–36
See also Success criteria
Internal dialogues in self-regulation, 156 (table)
Internal feedback, 181
Interpretation, evidence
always for individuals students, 80
implementation principles for formative assessment and, 250
See also Evidence
Involvement, student. *See* Student ownership and involvement

Keeley, P., 172, 173

Learning intentions, 7–8
  additional key characteristics of,
      31–36
  aligned to success criteria, 34–35
  aligning evidence to, 69–71
  characteristics of, 27–36
  clarification of goals using, 158–160
  creating, 36–58
  defined, 24–27
  and demonstrating reaching the
      learning, 46–52
  focused on learning, not activities,
      27–31
  focusing lesson on highest-priority
      learning for that lesson, 31–32,
      40–46
  helping students learn about, 167–175
  implementation indicators, 277–281
  implementation principles for formative
      assessment and, 249
  lesson planning with focus on, 38–52
  revisiting, 174–175
  teacher role in sharing and using, 53–58
  teachers supporting students' use of,
      56–58
  teacher use of, 53–55
  using unit progression to write,
      208–209
  wording of, 44–46
  written to be understandable by
      students, 35–36
  See also Success criteria
Learning materials, 240
Learning progressions, 11, 191–192, 277
  defined, 192–198
  developing student ownership and
      involvement using, 213–214
  helping to diagnose student difficulties,
      210–213
  implementation principles for formative
      assessment and, 249
  providing mathematics background
      information, 198–204
  used in planning instruction, 204–209
Learning resources, 60, 106, 150, 189,
      215–216, 244–245, 268, 275
Lesson planning
  choosing a responsive action and, 90–99
  considering the students in, 52
  demonstrating reaching the learning
      and, 46–52
  for eliciting evidence, 72–80
  for formative feedback, 130–134
  with learning intention and success
      criteria in mind, 38–52
  learning progressions used in, 204–209
  lesson-level planning, 274
  for most important learning, 40–46
  teacher role during lessons and, 53–58
  wording in, 44–46

Management of evidence information,
      100–101
Materials
  evidence-gathering, 241
  learning, 240
  resources on classroom, 62, 108, 151,
      190, 245, 275–276
  self-assessment, 241–242
Mathematics background information
  dependent and independent
      variables, 127
  expressions for figure patterns, 69
  proportional relationships in tables and
      graphs, 88–89
  using learning progressions to provide,
      198–204
*Mathematics Curriculum Topic Study:
      Bridging the Gap Between Standards and
      Practice*, 204
*Mathematics Formative Assessment:
      75 Practical Strategies for Linking
      Assessment, Instruction, and Learning*,
      172, 173
Math tasks and instructional environment,
      234–236
Misalignment of tasks with evidence,
      73–76
Model response
  "cutting corners" on creating,
      132–134
  planning, 130–132
Most important learning, 31–32, 40–46
Moving Learning Forward bulletin board,
      182–183

National Research Council, 29
National Science Foundation, 2
No-hands-up rule, 229
Nonlinear process of self-regulation,
      183–187

Ongoing teaching and learning, 6
Ownership and involvement, student,
      10–11, 154–156
  helping students develop, 166–183
  helping students learn about evidence
      and, 175–178
  helping students learn about formative
      feedback and, 178–181
  helping students learn about taking
      action and, 181–183
  implementation indicators, 279–281
  implementation principles for formative
      assessment and, 251–253
  learning intentions and success criteria
      to clarify goals for, 158–160,
      167–175
  learning to determine next steps and,
      162–164
  self-assessment and, 160–162, 175–178

self-regulation as nonlinear process and, 183–187

taking action in, 165–166

using learning progressions to help develop, 213–214

using success criteria, evidence, and formative feedback to self-assess in, 160–162

what students need to learn for, 156–166

Participation, student, 102–103
Patterns, expressions for figure, 69
Peers, 164
    as resources, 242–243
Physical environment, 219
    evidence-gathering materials in, 241
    keeping resources available, 237–240
    learning materials in, 240
    self-assessment materials in, 241–242
Pitfalls with success criteria, 50–52
Planning, lesson
    choosing a responsive action and, 90–99
    considering the students in, 52
    demonstrating reaching the learning and, 46–52
    for eliciting evidence, 72–80
    for formative feedback, 130–134
    with learning intention and success criteria in mind, 38–52
    learning progressions used in, 204–209
    lesson-level planning, 274
    for most important learning, 40–46
    teacher role during lessons and, 53–58
    wording in, 44–46
Planning resources, 61, 107, 150, 189, 216, 269, 274
Posters, 151, 190, 276
Praise, 115
Progressions, learning, 11, 191–192, 277
    defined, 192–198
    developing student ownership and involvement using, 213–214
    helping to diagnose student difficulties, 210–213
    implementation principles for formative assessment and, 249
    providing mathematics background information, 198–204
    used in planning instruction, 204–209
Proportional relationships in tables and graphs, 88–89
Providing Formative Feedback to Guide and Move Forward, 172

Questions
    curiosity and, 230–231
    hinge-point, 75–76
    during lessons, 83
    used to jump to instruction, 84–86

Reference resources, 60–61, 107, 150, 189, 216, 245, 269, 272–275
Reflect Aloud, 174
Resources
    classroom, 61–62, 107–108, 150–151, 189–190, 274–276
    classroom materials, 62, 108, 151, 190, 275–276
    learning, 60, 106, 150, 189, 215–216, 244–245, 268, 271–275
    peers as, 242–243
    planning, 61, 107, 150, 189, 216, 269, 274
    reference, 60–61, 107, 150, 189, 216, 245, 269, 272–275
Response cards, 75, 76
Responsive actions, 90–99
    formative feedback as, 128–130
    helping students learn about, 181–183
    by students receiving formative feedback, 147–148
Revisiting learning intentions, 174–175
Risk-taking and intellectual safety, 223–229

Sadler, R., 3
Safety, intellectual
    curiosity and, 222–223
    risk-taking and, 223–229
Self-assessment
    materials, 241–242
    student ownership and, 160–162, 175–178, 190
Self-assessment, student, 160–162, 175–178
Self-monitoring by students, 103–105
Self-regulation, 2, 3–4
    evidence and, 64–65, 175–178
    formative feedback and, 113, 114 (figure), 143–144, 178–181
    internal dialogues, 156 (table)
    learning intentions and success criteria to clarify goals for, 158–160
    learning to determine next steps and, 162–164
    as nonlinear process, 183–187
    poster, 190
    self-monitoring and, 103–105
    in student ownership and involvement, 154–156
    students taking action in, 165–166
    taking action and, 181–183
    use of learning intentions and success criteria in, 56–58
    using formative assessment practices to develop skills in, 5–11
    using success criteria, evidence, and formative feedback to self-assess in, 160–162
    what students need to learn for, 156–166

Social/cultural environment, 218
  promoting intellectual safety and
    curiosity, 222–223
  putting together norms and
    recommendations for, 233
Standards for Mathematical Practice, 30
  See also Common Core State Standards
Student ownership and involvement,
    10–11, 154–156
  helping students develop, 166–183
  helping students learn about evidence
    and, 175–178
  helping students learn about formative
    feedback and, 178–181
  helping students learn about taking
    action and, 181–183
  implementation indicators, 279–281
  implementation principles for formative
    assessment and, 251–253
  learning intentions and success criteria
    to clarify goals for, 158–160, 167–175
  learning to determine next steps and,
    162–164
  self-assessment and, 160–162, 175–178
  self-regulation as nonlinear process and,
    183–187
  taking action in, 165–166
  using learning progressions to help
    develop, 213–214
  using success criteria, evidence, and
    formative feedback to self-assess in,
    160–162
  what students need to learn for, 156–166
Students
  becoming more engaged in mathematics
    learning, 1–2
  collaboration by, 231–232
  considered during lesson planning, 52
  difficulties diagnosed using learning
    progressions, 210–213
  formative feedback that is usable by,
    120–122
  interpreting evidence always for
    individual, 80
  participation, 102–103
  as peer resources, 242–243
  responsibility for doing the learning,
    236–237
  role in evidence elicitation, 102–105
  role in formative assessment, 15–16
  role in formative feedback, 143–148
  self-monitoring by, 103–105
  as self-regulating learners, 2, 3–4
  summary cards, 62, 108, 275–276
  taught how to participate in formative
    assessment, 18–20
  teachers supporting use of learning
    intentions and success criteria by,
    56–58
  See also Self-regulation

Success criteria, 7–8
  additional key characteristics of, 31–36
  aligned to learning intention, 34–35
  aligning evidence to, 69–71
  characteristics of, 27–36
  clarification of goals using, 158–160
  common pitfalls to avoid with, 50–52
  creating, 36–58
  defined, 24–27
  demonstrating reaching the learning,
    46–52
  evidence-gathering saved for end of
    class and, 76–80
  formative feedback providing guidance
    on what student needs to do next to
    meet, 118–119, 136–141
  formative feedback referencing extent of
    meeting, 117
  helping students learn about, 167–175
  implementation indicators, 277–281
  implementation principles for formative
    assessment and, 249
  as indication of reaching learning
    intention, 33–34
  interpreting evidence of meeting, 86–90
  lesson planning with focus on, 38–52
  not always "leveled," 51
  resources, 275
  self-assessment and student ownership,
    160–162
  sounding like the lesson agenda, 50
  tangible or observable examples, 32–33
  teacher role in sharing and using, 53–58
  teachers supporting students' use of,
    56–58
  as too specific, 50
  turning into a long list, 50–51
  wording of, 44–46
  written to be understandable by
    students, 35–36
  See also Learning intentions
Summary cards, 108, 151, 190, 245, 272–273
Supporting aspects of formative
    assessment, 11
Sustainability of formative assessment
    efforts, 254–266

Taking Stock to Clarify and Consolidate,
    172, 178
Teachers
  benefiting from working collaboratively,
    258–260
  comparing formative assessment
    practices to teaching practices,
    16–17
  focusing on own learning first, 255–257
  role during lessons and eliciting
    evidence, 53–58
  role during lessons and formative
    feedback, 134–143

role in formative assessment, 15–16
supporting students' use of learning intentions and success criteria, 56–58
taking time for formative assessment, 257–258
Templates, 276
Think/Pair/Share Strategy, 169
3-Read Strategy, 169
Timers, countdown, 229
Tobey, C., 172, 173

Understanding of formative feedback, 144–146
Unit-level planning, 274

Unit progressions, 205
for clarifying concepts, 206–208
to write learning intentions, 208–209

Variables, dependent and independent, 127

Web-based interactives, 271–272
Whole-group feedback, 123–128
Wiggins, G., 121–122
William, D., 12, 21, 169, 236
Wording, 44–46

X Marks the Spot, 173, 178

Corwin is committed to improving education for all learners by publishing books and other professional development resources for those serving the field of PreK–12 education. By providing practical, hands-on materials, Corwin continues to carry out the promise of its motto: **"Helping Educators Do Their Work Better."**